EUROPE AND THE POST-YUGOSLAV SPACE

Europe and the Post-Yugoslav Space

Edited by

BRANISLAV RADELJIĆ
University of East London, UK

ASHGATE

Published by
Ashgate Publishing Limited
Wey Court East
Union Road
Farnham
Surrey, GU9 7PT
England

Ashgate Publishing Company
110 Cherry Street
Suite 3-1
Burlington, VT 05401-3818
USA

www.ashgate.com

British Library Cataloguing in Publication Data
Europe and the post-Yugoslav space.
1. European Union--Former Yugoslav republics. 2. Political culture--Former Yugoslav republics.
I. Radeljić, Branislav.
327.4'97'04-dc23

The Library of Congress has cataloged the printed edition as follows:
Radeljić, Branislav
Europe and the post-Yugoslav space : / by Branislav Radeljić.
pages cm
Includes bibliographical references and index.
ISBN 978-1-4094-5390-1 (hardback) -- ISBN 978-1-4094-5391-8 (ebook) -- ISBN 978-1-4724-0456-5 (epub) 1. European Union--Yugoslavia. 2. Yugoslavia--European Union. 3. European Union countries--Economic integration. I. Title.
HC240.25.Y8R33 2013
341.242'209497--dc23

2012048048

ISBN 9781409453901 (hbk)
ISBN 9781409453918 (ebk – PDF)
ISBN 9781472404565 (ebk – ePUB)

MIX
Paper from responsible sources
FSC® C018575

Printed and bound in Great Britain
by MPG PRINTGROUP

Contents

List of Tables

Notes on Contributors

Jelena Džankić is a Jean Monnet Fellow at the European University Institute in Florence. She holds a PhD from the Faculty of Politics, Psychology and Social Sciences at the University of Cambridge (New Hall College). Her academic interests fall within the fields of the transition of South East European countries, Europeanization, and the politics of identity and citizenship. She has published several articles on the politics of the Balkan region, and since 2009 has been the country expert on Montenegro at the European Union Democracy Observatory on Citizenship at the Robert Schuman Centre for Advanced Studies.

Simonida Kačarska is a PhD candidate in the School of Politics and International Studies at the University of Leeds in the UK. She holds a BA in Political Science-International Relations and European Studies from the American University in Bulgaria and an MA in European Politics from the University of Sussex in the UK. She has both practitioner and research experience in the area of the political transformation of the Western Balkans and the European integration of this region. She has published in both Macedonian and international journals on the topics of democratization of the Western Balkans and its European integration.

Damjan Mandelc is an assistant professor at the Department of Sociology, Faculty of Arts, University of Ljubljana, Slovenia. His research interests include the sociology of nation and nationalism, the sociology of globalization, the role of national identity and nation-state in the processes of integration, citizenship and European Union politics. He took an active part in the projects on 'The Role of Mass Media in the Formation of the Slovene European Consciousness' and 'Small Nations and the European Public Sphere'. Currently he is involved in the research programme entitled 'Problems of Autonomy and Identities in the Global Age'. In addition, his work has been published both in Slovenia and abroad.

Leandrit I. Mehmeti is completing his PhD at the University of South Australia. He is interested in the integration of the Western Balkans in the European Union and the collective security challenges. He teaches the subjects of the EU and World Politics and Global Societies at the University of South Australia. Previously, he worked with the OSCE Mission in Kosovo as a National Political Affairs Officer, where he had the opportunity to work closely with political parties, civil administration and civil society in the Balkan region.

Ljubomir Mikić is the head of Service for the Political System and National Minorities in the Ministry of Public Administration of Croatia. He has dealt professionally with refugee and minority rights issues in Croatia and the region, in both the public and non-governmental sectors. He has also authored and co-authored a number of analyses, reports and policy papers on the state of human and minority rights in Croatia.

Antonija Petričušić is a research and teaching assistant at the Department of Sociology at the Faculty of Law, University of Zagreb, Croatia. She is pursuing her PhD studies in constitutional law at the Karl-Franzens University of Graz, Austria. She has authored a number of articles and studies dealing with human rights and protection of national minorities in South East Europe. In addition to her academic work, she served as a member of the Croatian negotiations team in the chapter dealing with judiciary and fundamental rights.

Branislav Radeljić is a senior lecturer in international politics at the University of East London's School of Law and Social Sciences. His main research interests focus on the study of European Union politics and Eastern Europe. He is the author of *Europe and the Collapse of Yugoslavia: The Role of Non-State Actors and European Diplomacy* (London and New York: I.B. Tauris, 2012). In addition, Branislav has also examined and written about the presence of Islam in the EU and its impact on European identity and EU policy making.

Marko Stojić is a PhD candidate and an associate tutor at the Sussex European Institute, University of Sussex. He has a BA from the Faculty of Political Science, University of Belgrade, and an MA in International Relations and European Studies from the Central European University, Budapest. His research interests focus on the study of political parties and party systems in the Western Balkans, and include the domestic politics of European integration, the attitudes of political parties towards the EU and party-based Euroscepticism in the former Yugoslavia.

Emir Vajzović is a lecturer in democracy, human rights and EU integrations at the Faculty of Political Science, University of Sarajevo, and programme director of the Institute for Social Sciences Research of the Faculty of Political Science, Sarajevo. His research interests include media democracy, EU integration processes and human rights. He has a BA in Journalism and Communications from the University of Sarajevo, an MA in Democracy and Human Rights from the University of Bologna and University of Sarajevo, and is completing a PhD in Political Sciences at the University of Sarajevo.

Preface and Acknowledgements

The idea for this volume originated in 2011 while I was completing the last section of a manuscript on the relations between the then European Community and the Socialist Federal Republic of Yugoslavia. Given that it addressed the problematic period lasting from 1968, when official relations between the two parties were established, to 1992, when the Community recognized the disintegration of Yugoslavia, the new academic inquiry concerned Europe and the post-Yugoslav space. As has already been the case in numerous public debates and scholarly accounts, the term 'post-Yugoslav' raises immediate questions about contributors' intentions. Here, it refers to a well-known territory and the new states created out of the Yugoslav federation, their progress towards European Union membership and the often challenging regional cooperation amongst them. The selection of a term that includes Yugoslavia should not be interpreted here as an ambition to resurrect a state that ceased to exist two decades ago, but as an analytical and reflective tool capable of distinguishing between the good and, more importantly due to the outcome, bad aspects that characterized the Yugoslav state.

This edited volume brings together a diverse group of emerging scholars originally coming from the former Yugoslavia and its respective post-Yugoslav states. More precisely, each state is addressed by an author coming from that particular state but who has spent some time abroad (for educational, project or workshop purposes) and thus has had an opportunity to observe the developments from a distance. It is hoped that this will reduce the amount of bias that has dominated debates about pre- and post-Yugoslav dynamics over the years; indeed, various authors, both local and international, have continuously tried to convince us who the bad guys were, while at the same time sparing a number of other aspects from serious criticism. Finally, the selection of topics has to do with the authors' own perceptions of the issues they found of great relevance when thinking about the role played by the European Union in the post-Yugoslav states.

Chapter 1 provides an overview of European involvement in the post-Yugoslav space by looking mainly at the progress of individual states towards their integration in the European Union. By examining official EU statements, the chapter shows that the speed of democratization and Europeanization has been rather uneven, and something that can work in one of the states does not necessarily have to be a model for the others. The fact that all of them have experienced diverse difficulties sustains the much-debated topic about (in)appropriate strategies. Thus, as states are expected to implement necessary reforms and close relevant chapters – perceived as absolute prerequisites for EU membership – this chapter seeks to identify the aspects that have affected the whole process (both positively and negatively) and who is to bear responsibility, the EU aspirants, the Union itself, or both.

In Chapter 2, Damjan Mandelc argues that Slovenia intentionally distanced itself from the post-Yugoslav space in order to secure its status in the EU and only recently has begun to intensify its presence by openly supporting and getting involved in regional cooperation among the rest of the former Yugoslavia. However, in some other areas, as nicely illustrated, the Slovenes (including new, post-Yugoslav, generations) have always tried to stay in touch with other Yugoslav peoples. Such observations should not be interpreted as the author's wish to see a new Yugoslavia emerge in the future, but as an indication of potential. As noted, what could help is the establishment of stronger links resulting in joint projects and more profound cooperation among the post-Yugoslav states, hopefully within the EU.

Still, although often presented as a successful case and a model, Slovenia could also be criticized for some of its domestic policies during the processes of Europeanization and democratization. Here, Mandelc confirms his impartial point of view, especially when discussing some problematic and, moreover, sensitive issues, such as discrimination in post-independence Slovenia, and the mainly inadequate treatment of minorities and citizenship concerns.

Chapter 3 is about Slovenia's neighbour, Croatia. For Antonija Petričušić and Ljubomir Mikić, by stressing the relevance of minority rights and, more importantly, by including them in its accession criteria, the EU is likely to enhance the position of national minorities in Croatia as well as to contribute to peaceful relations in the region. In order to elaborate on and sustain such a conclusion, the authors rely on rational choice theory, EU-released progress reports and other available data. They say that, given the initial security obstacles, only a small number of refugees of Serb origin decided to return to Croatia, although this number was higher back in the late 1990s than today.

The EU's decision to make the protection of minorities a 'soft' pre-accession political criterion for Croatia (and other EU-aspiring countries) and the regional initiative known as the Sarajevo Process (gathering governments of Bosnia and Herzegovina, Croatia, Montenegro and Serbia) under international community guidance, led to an increase in the number of returnees across the region and may even have brought the population displacement to an end. In addition, such moves generated continuous support from individual governments. However, as clearly underlined, the disappointing economic performance and high unemployment in the area continue to represent the main obstacle to return.

In Chapter 4, Simonida Kačarska addresses the role of the EU (as an 'active player' and 'framework for integration') via the political conditionality in Macedonia in the period 1997–2011. Although the country was the first in the region to sign the Stabilisation and Association Agreement and the second to become an EU candidate, since 2006, Macedonia's accession process has slowed down. For example, the author notes that the dispute over Macedonia's constitutional name further complicated the transition of the EU from being an active player to providing a framework for integration, thus questioning the legitimacy of the EU involvement.

However, the Brussels administration decided to assist the country with various initiatives, such as the visa liberalization process and high-level accession dialogue. The chapter combines concepts from the Europeanization and conditionality literature with studies on the EU's performance in conflict settlement and resolution. In addition, it benefits from extensive use of primary sources, obtained from numerous interviews with stakeholders, conducted in Brussels and Skopje. The fresh conclusions reached are even more relevant given the enlargement plans outlined by the European Commission in 2011.

Discussion about the changing perception and role of the EU in Montenegro dominates Chapter 5. Jelena Džankić starts by looking at the first years after the disintegration of Yugoslavia (when Montenegro was a constituent state of the Federal Republic of Yugoslavia), then moves on to look at the relevance of the split of the ruling party in Montenegro and the transformation of the then state into the State Union of Serbia and Montenegro, and concludes by examining the EU's role in the post-independence period.

In order to prove the main arguments, the author combines the theory on European foreign policy with the vast literature on Europeanization, an approach that is complemented by empirical data acquired through the employment of qualitative methodology (documentary survey and context analysis). Put together, she comes to a set of insightful conclusions about each of the stages of the EU–Montenegro relationship. Of course, the final, post-2006, stage is all about the EU's conditionality and the process of Europeanization of Montenegro.

Marko Stojić elaborates on the political scene in Serbia in Chapter 6. Interestingly, in post-2000 Serbia, some of the nationalists became supporters of Serbia's membership of the EU, while some democrats who had actively supported the overthrow of Slobodan Milošević came out against membership. In order to explain the causes and consequences of Euroscepticism among Serbian political elites and its relevance for the political system, the chapter benefits from political parties' statements and records of parliamentary voting on key European issues, as well as interviews conducted with senior party officials from all relevant parties.

Accordingly, the chapter offers various conclusions. For example, the political parties in Serbia do not know much about the EU and thus avoid serious discussions about Serbia's potential membership. Generally, liberal and modernist parties tend to be pro-European, while traditional and national parties are rather Eurosceptic. Still, some representatives have adopted strategically motivated attitudes towards Europe and thus tend to change their position with regard to Serbia's EU membership. Also, it should be remembered that the EU is still mainly perceived and discussed based on its policies towards the former Yugoslavia in the early 1990s and towards Serbia (mainly vis-à-vis the status of Kosovo).

In Chapter 7, Emir Vajzović examines Bosnia and Herzegovina's European perspective. He argues that the EU has struggled to find the best way to use its carrot-and-stick approach in this post-Yugoslav state. As noted, the EU has done well in some fields (including bringing war to an end, instigating the stability process and reconstruction, financial assistance, etc.), while in some others (including the failed police reform) its involvement has been quite problematic. For example, the criteria and conditions for further progress in the Stabilisation and Association Process have changed and remain unclear. The EU has also been imprecise about constitutional reform, maintaining that constitutional changes are not a requirement, but they are necessary.

To complicate the whole situation further, political elites in Bosnia and Herzegovina, although in favour of EU accession, seem not to be ready to give up their power for the sake of accession to the EU without a proper carrot-and-stick arrangement. However, the author wants to believe that the main driving force rests with the local people, who, while managing to overcome the obstacles from the past, understand the political mosaic and are ready to hold their representatives responsible for actions taken.

In Chapter 8, Leandrit I. Mehmeti assesses the role of international actors in the democratization process in Kosovo. He argues that Kosovo has made significant progress with regard to institution building, even though the whole process has faced various challenges, especially after the 1999 war. In this process, the EU is making every effort to bring the Kosovo and Serbian authorities together in order to find a sustainable solution. According to the author, the Serbian leadership tends to ignore the events preceding and requiring the 1999 NATO-led intervention and thus continues to insist on the Serbian right to the province. On the other hand, the Kosovo Albanians tend to deny the humiliation and revenge perpetrated against the remaining Serbs.

However, the chapter notes that although the United States of America and Russia seem to be the main international players in the province, it is the EU that holds the key to peace in the region by being actively involved in the processes of democratization and economic development. As underlined, while challenges are indeed great, given the diametrically opposing views about Kosovo's status, it is the integration into the EU that should be the main objective of the Brussels administration as well as the Serbian and Kosovo Albanian sides.

Finally, the last chapter of the volume will round off the debate by putting it into a broader context. While aware of the events characterizing and, most importantly, affecting the expected progress of the post-Yugoslav space, this section will outline some of the dominant challenges and possible responses deriving from both the European Union and the local, EU-aspiring, states. The contributors' decisions to tackle a range of diverse aspects, some of which could also be found in multiple post-Yugoslav states, confirms the complexity of the topic on its own and thus the importance of durable solutions.

Indeed, the authors deserve a large portion of the editor's appreciation, as without their ideas and hard work this volume could not have been possible. Some segments of their research and fresh conclusions have already been presented and appreciated at various international conferences and workshops. Apart from the authors, it is a pleasure to thank the Ashgate team, especially Rob Sorsby, Senior

Commissioning Editor for International Politics, and Kayleigh Huelin, Assistant Desk Editor, for their advice and support from the very beginning to the finishing line. Also, it is important to thank the anonymous reviewers for providing constructive feedback and suggestions on earlier versions of the chapters as well as the volume as a whole. Finally, it is imperative to acknowledge that working on this book was substantially supported by a grant received from the University of East London's School of Law and Social Sciences.

List of Abbreviations

ASSC	Areas of Special State Concern
CARDS	Community Assistance for Reconstruction, Development and Stabilisation
CCI	Centre of Civil Initiatives
CEE	Central and Eastern Europe
CEEC(s)	Central and Eastern European Country(ies)
CFSP	Common Foreign and Security Policy
CIN	Center for Investigative Reporting
CSD	Center for the Study of Democracy
DOS	Democratic Opposition of Serbia
DPA	Dayton Peace Agreement
DPMNE	Democratic Party for Macedonian National Unity
DPS	Democratic Party of Socialists (Montenegro)
DSS	Democratic Party of Serbia
DUI	Democratic Union for Integration (Macedonia)
EC	European Community
EPC	European Political Cooperation
ERW	Explosive remnants of war
ESI	European Stability Initiative
EU	European Union
EUFOR	European Union Force
EULEX	European Union Rule of Law Mission
EUPM	European Union Police Mission
EUSR	European Union Special Representative
FPÖ	Freedom Party of Austria
FRY	Federal Republic of Yugoslavia
HDZ	Croatian Democratic Union
ICG	International Crisis Group
ICJ	International Court of Justice
ICTY	International Criminal Tribunal for the former Yugoslavia
IDP(s)	Internally displaced person(s)
JNA	Yugoslav People's Army
KFOR	Kosovo Force
KIPRED	Kosovar Institute for Policy Research and Development
KLA	Kosovo Liberation Army
KQZ	Central Election Commission
LCY	League of Communists of Yugoslavia

LDK	Democratic League of Kosovo
NATO	North Atlantic Treaty Organization
NS	People's Party (Montenegro)
NS	New Serbia
OECD	Organisation for Economic Cooperation and Development
OFA	Ohrid Framework Agreement
OHR	Office of the High Representative
OSCE	Organization for Security and Cooperation in Europe
PDP	Party for Democratic Prosperity (Macedonia)
PIC	Peace Implementation Council
RS	Republika Srpska (Serb Republic)
SAA	Stabilisation and Association Agreement
SAP	Stabilisation and Association Process
SCG	Serbia and Montenegro
SDP	Social Democratic Party (Montenegro)
SEE	South Eastern Europe
SEIO	European Integration Office (Serbia)
SFRY	Socialist Federal Republic of Yugoslavia
SNS	Serbian Progressive Party
SPS	Socialist Party of Serbia
SRS	Serbian Radical Party
UN	United Nations
UNHCR	United Nations High Commissioner for Refugees
UNMIK	United Nations Mission in Kosovo
UNODC	United Nations Office on Drugs and Crimes
UNPREDEP	United Nations Preventive Deployment Force
UNPROFOR	United Nations Protection Force
US	United States of America
USY	United States of Yugoslavia
VMRO	Internal Macedonian Revolutionary Organization
WB	Western Balkans
WWII	Second World War

Chapter 1

Europe and the Post-Yugoslav Space: From Intervention to Integration

Branislav Radeljić

The Socialist Federal Republic of Yugoslavia (SFRY) ceased to exist on 15 January 1992 when the Member States of the then European Community (EC) and the present European Union (EU) agreed to grant independence to all the Yugoslav republics that, in their view, satisfied certain conditions with regard to democracy, the rule of law, the rights of ethnic and national groups, the inviolability of borders, security and regional stability, and so on. The decision to recognize Slovenia and Croatia on 15 January, followed by recognition of Bosnia and Herzegovina on 7 April 1992 and the Former Yugoslav Republic of Macedonia in 1995 was the result of the courageous statement of Jacques Poos, European Community representative, delivered at the EC's Foreign Affairs Council of Ministers in June 1991, that 'the hour of Europe' had come and '[i]f one problem can be solved by the Europeans, it is the Yugoslav problem. This is a European country and it is not up to the Americans. It is not up to anyone else' (Poos cited in Smith 1996: 1).

This chapter offers an overview of the dynamics characterizing European involvement in the (post-)Yugoslav space, through intervention and then integration. While in some cases European policies managed to provide an opportunity for a straightforward path towards EU membership (in the case of Slovenia and, to a significantly lesser extent, Croatia), in some others European involvement has been much more problematic and uncertain, without clear strategies and responses from the locals, thus risking creating further divisions between the EU and the rest of the Yugoslav space.

From Intervention to Integration

The outbreak of the Yugoslav state crisis, the consequent wars, European Union policies and the future of the post-Yugoslav space have received enormous attention in the literature. Questions focusing on what went wrong with the SFRY, why its disintegration was so violent, whether the adopted policies were the right ones and what should be done in order to avoid similar scenarios in the future have emerged over time, generating diverse interpretations and conclusions. In fact, various literature reviews have tried to classify the existing arguments, based on their accountability and relevance for deeper understanding of Yugoslavia's collapse, and to identify some additional aspects in need of examination (Dragović-Soso 2007, Gow 1997, Jović 2001, Kent 1997, Nakarada 2008, Radeljić 2012, Ramet 2005, Stokes et al. 1996).

The first European intervention in Yugoslavia resulted in the collapse of the Yugoslav federation. The decision to recognize Slovenia and, more importantly, Croatia, meant that the war was then transferred to Bosnia and Herzegovina. With regard to the European Union's initial approach, one author assessed the performance of its leaders as highly immature:

> However much the EC leaders liked to think in the long term – charting the future of a continent like master mariners setting forth on the great ocean – the leaders of the Twelve proved remarkably lacking in foresight when it came to anticipating events in the Balkans. Not only were they taken by surprise by the flare-up of fighting in Slovenia, but they refused to be distracted from each fresh outbreak of violence to consider the big picture. For instance, already in July 1991 the government of Bosnia and Herzegovina was so alarmed by the scale of the fighting along its north-western borders inside Croatia and the involvement of JNA and militias based in Bosnia that it asked the EC to engage in preventive diplomacy … The Twelve displayed sovereign contempt for Sarajevo's attempt to distract them from their central preoccupation … Buffeted by the high tide of history, the Twelve continued according to routines better suited to calmer waters. (Almond 1994: 242–3)

Simultaneously with the continuation of the fighting in Yugoslavia, the 1993 Copenhagen European Council addressed the relations between the Community and the countries of Central and Eastern Europe and endorsed a set of criteria to be satisfied if they wanted to become members of the EU (EU Council 1993).[1] This was a clear encouragement and an open call for a new region to try to integrate into European structures and it was logical that such an experience would strongly affect the post-Yugoslav space as well. Consequently, in February 1996, the EU adopted a Regional Approach to the countries of South East Europe and, in April 1997, the Council established the political and economic conditions to be fulfilled by these countries. Later, following the second European intervention

1 The European Union has defined the conditions for enlargement as follows: 'The so-called "Copenhagen criteria", set out in December 1993 by the European Council in Copenhagen, require a candidate country to have: stable institutions that guarantee democracy, the rule of law, human rights and respect for and protection of minorities; a functioning market economy, as well as the ability to cope with the pressure of competition and the market forces at work inside the Union; the ability to assume the obligations of membership, in particular adherence to the objectives of political, economic and monetary union. In 1995 the Madrid European Council further clarified that a candidate country must also be able to put the EU rules and procedures into effect' (available at: http://ec.europa.eu/justice/international-relations/enlargement/index_en.htm).

in the post-SFRY space (this time as part of NATO's action to save Kosovo), the launch of the Stability Pact for South Eastern Europe in 1999 (replaced by the Regional Cooperation Council in 2008), the Reconstruction Agency in 2000, and the Stabilisation and Association Process in 2001, were to serve as the main instruments to consolidate regional stability and cohesion.

Since then, European Union officials have continued to use various occasions to confirm their commitment to the post-Yugoslav states or, as has been heard more often, to the Western Balkans, an EU neologism that excludes Slovenia (for understandable reasons) and includes Albania. For example, in Thessaloniki in 2003, they underlined that '[t]he future of the Balkans is within the European Union', but of course, such a direction would be conditional on the fulfilment of the same requirements that applied to the Central European states (EU Council 2003). Similarly, in 2008, the Europeans confirmed their standpoint: 'The future of the Western Balkans lies in the EU. The EU stresses the importance of peace, stability and security in this part of Europe, and welcomes all efforts of the Western Balkan countries to come closer to the EU, meeting the necessary conditions. The Western Balkans have the potential to accelerate their course towards eventual EU membership, provided they pursue the path of reform and reconciliation, and meet the necessary conditions. The EU will assist them in this endeavor' (European Commission 2008a: 21).

It can be argued that the EU has heavily relied on enlargement as its most celebrated foreign policy tool to penetrate the post-Yugoslav space. While in some cases such an approach has produced significant results, in others it has often been stuck between the EU's conditionality clauses and the applicants' willingness to comply. As witnessed, not all post-Yugoslav states have perceived and welcomed the processes of democratization and Europeanization in the same way, although fully aware of possible benefits. While some of them have progressed from signing the Stabilisation and Association Agreement (SAA) to joining the EU, via various adjustments and liberalization plans, others have seriously struggled to pursue the necessary reforms and modify their systems in order to secure candidate status and eventual membership in the EU.

The Progress of the Post-Yugoslav States

Slovenia

Back in the late 1970s the European Parliament hosted some lengthy debates about the Yugoslav federation that clearly confirmed the European Community's appreciation for Slovenia's economic advancement. Due to its location and human capital, the republic might have become an 'experimental zone of big interest for the Community' to test whether its new technical norms would be applicable across other Yugoslav areas in the future (European Parliament 1977). The economic as well as other links between the Europeans and the Slovenes were further strengthened throughout the 1980s. As one study correctly puts it, '[f]avorable economic prospects and the positive inducements of joining Europe helped to shape a pluralist orientation on the part of elites and masses and, ultimately, to crate and reinforce democratic attitudes in the long run' (Boduszyński 2010: 116).

The growing disagreement with the Yugoslav authorities culminated in 1989 when the Slovene Parliament amended the republic's constitution confirming Slovenia's right to secede from the SFRY and welcoming political pluralism. In April 1990, in the first free, multi-party, elections, Milan Kučan, a liberal communist advocating democratization and Europeanization of Slovenia, was elected as president. Eight months later, in a referendum on independence, a clear majority of Slovenes confirmed their dissatisfaction with the Yugoslav federation, its leadership and system. The proclamation of independence on 25 June 1991 and the consequent Ten-Day War, also referred to as the Slovenian Independence War, provided the Slovenes with an opportunity to insist even more on the importance of concepts such as democracy, freedom and human rights. Slovenia's pro-European leadership, while already supported by various state and non-state actors, perceived international recognition of independence as the only way forward. At this point, being pro-European meant being pro-democracy and pro-modernization, two aspects Yugoslavia was seriously struggling with.

Once independence had been achieved, a strong incentive for European Union accession gained its full relevance amongst Slovenia's authorities and the public: 'Slovenia's leaders relentlessly pursued the goal of EU integration so that, after a while, a seemingly unshakable social consensus existed on the imperative of EU membership to assure stability, democracy and prosperity in Slovenia. By the mid-1990s the public, for its part, was under no illusions about the drawbacks of membership' (Boduszyński 2010: 128). In fact, after President Kučan's visit to Brussels in March 1992, the Council authorized the European Commission to negotiate a Cooperation Agreement with Slovenia, a Financial Protocol and an Agreement on Transport (European Commission 1992: 284–5). Later, Slovenia signed the Association Agreement and applied for EU membership in 1996, which, in Kučan's view, confirmed that the country 'has gone a long way towards fulfilling the expectations of its citizens at the 1990 plebiscite and the declaration of independence ... that Slovenia would take its place in Euro-Atlantic institutions' (Kučan 1996).

According to the 1997 European Commission assessment of Slovenia's application for EU membership, the country's accession was 'seen as part of an historic process, in which the countries of Central and Eastern Europe overcome the division of the continent' (European Commission 1997a: 5). With regard to the fulfilment of the political criteria, Slovenia already 'present[ed] the characteristics of a democracy with stable institutions guaranteeing the rule of law, human rights and respect for and protection of minorities' (European Commission 1997a: 115). With regard to the economic criteria, the Commission described Slovenia as 'a functioning market economy' that 'should be able to cope with competitive pressure and market forces within the Union in the medium term, provided that rigidities in the economy are reduced' (European Commission 1997a). The report pointed out areas requiring improvement (administration, energy, environment, employment, internal market, social affairs), but still confidently recommended the opening of negotiations for Slovenia's EU membership.

The accession process was formally launched in March 1998 and the following reports acknowledged that Slovenia was taking the Copenhagen criteria seriously and that it actually 'demonstrated impressive progress across most sectors of the acquis' (European Commission 1999: 71). Even reforms that were initially lagging behind, such as in agricultural, environmental, judicial and public administration sectors, were pursued. Thus, the Europeans were pleased to see that Slovenes were committed to the previously established objectives and each of the following reports clearly expressed appreciation for their performance. In fact, the 2002 report concluded: 'Bearing in mind the progress achieved since the [1997] Opinion, the level of alignment and administrative capacity that Slovenia has achieved at this point in time, and its track record in implementing the commitments it has made in the negotiations, the Commission considers that Slovenia will be able to assume the obligations of membership in accordance with the envisaged timeframe' (European Commission 2002b: 121).

Slovenia joined the European Union and NATO in 2004 and the Euro area in 2007. In fact, the country's six-month presidency of the EU in 2008 was a clear confirmation of 'Slovenia's European credentials and its notable maturity on the international scene' (Gow and Carmichael 2010: 244). More importantly, the programme of the presidency, entitled *Si.nergija za Evropo/Si.nergy for Europe*, paid significant attention to the European future of the Western Balkans:

> Two key achievements of the Presidency were the completion of a network of Stabilisation and Association Agreements with all the countries of the region except Kosovo, and the start of dialogue on visa liberalisation and the presentation of roadmaps for visa liberalisation for all countries of the region (except Kosovo). The commitment was reaffirmed to strengthen and speed up the EU integration process of these countries and, building on the achievements to date, to draw up individual roadmaps for them. (EU Council 2008c: 18)

Following the presidency of the EU, the National Assembly of the Republic of Slovenia adopted a *Declaration on the Western Balkans* in 2010, confirming Slovenia's commitment to cooperation and development in the region, which would eventually become a full part of Euro-Atlantic structures. Such an explicit commitment has managed to convince the Brussels administration of Slovenia's readiness to encourage and strengthen regional cooperation in this part of Europe and thus consolidated the country's relevance for European politics. Equally important is the impression that the Slovenes themselves have reached the point where they are not exclusively perceived by some in the region as the ones who helped dismember the Yugoslav state but as those whose rapid progress represents the model to follow.

Croatia

The first multi-party elections in Croatia were held in April 1990 and won by the previously established Croatian Democratic Union. Its founder, Franjo Tudjman, a great advocate of Croatian interests, became president. The newly introduced constitution, apart from providing the republic with a right to secede from the SFRY, presented Croatia as the statehood of Croats. As a response to this, the ethnic Serbs decided to establish their own separatist entity within Croatia, the Republic of Serbian Krajina. Following the referendum on independence of Croatia organized in May 1991, in which a large majority of voters supported secession, and the consequent declaration of independence in June 1991, the tensions between the local Serbs and Croats increased and led to the outbreak of war.

European Community involvement consisted of criticizing the Serbian side: 'It is a deeply misguided policy on the part of the Serbian irregulars to try to solve the problems they expect to encounter in a new constitutional order through military means. It is even more disconcerting that it can no longer be denied that elements of the Yugoslav People's Army are lending their active support to the Serbian side' (EPC 1991a). The main objective of the EC *Conference of Yugoslavia*, launched in The Hague in September under the chairmanship of Lord Carrington, former British Foreign Minister, was to bring peace to all in Yugoslavia and discuss the future of the federation and its peoples. However, his efforts and those of Cyrus Vance, the former US Secretary of State, were of little help. As one of numerous declarations on the SFRY in this period observed, the Europeans were 'particularly disturbed by reports of continued attacks on civilian targets by elements of the federal armed forces and by both Serbian and Croatian irregular forces. The continuation of military activities in Croatia threatens to extend the armed confrontation to other regions of Yugoslavia' (EPC 1991b). The Europeans continued to condemn the escalation of ethnic intolerance and aggressive nationalism and in December they drafted the *Guidelines on the Recognition of New States in Eastern Europe and in the Soviet Union*.

The European decision to recognize the independence of Slovenia and Croatia on 15 January 1992 eased the position of the Slovenes, but not that of the Croats, Bosniaks and Serbs. Officials were naïve to believe that the recognition and consequent efforts of the EC monitoring mission and United Nations (UN) peacekeeping forces, together with the active participation of all parties concerned, were going to generate almost immediate settlement of the crisis. However, parallel to the fighting, the new Croatian authorities cultivated their relations with the West, mainly to secure support for their actions. The Tudjman regime, while relying on significant support from the Catholic Church (both in Croatia and abroad), diaspora communities and the international media, tried to convince the West of Croatia's commitment to liberal and pro-Western orientation, and this is why some governments did not object to his nationalistic policies, which at various points were reminiscent of the ones pursued by the Serbian leadership.

The end of the Croatian War of Independence in 1995 was accompanied by disappointing economic performance, growing unemployment rates, external criticism and pressure, and thus general international isolation of Croatia from Euro-Atlantic integration. Based on the constitution, regional cooperation was considered illegal: 'It is prohibited to initiate any procedure for the association of the Republic of Croatia into alliances with other states if such association leads, or might lead, to a renewal of a South Slav state community or to any Balkan state form of any kind' (Article 141 (2)).

Understandably, even the strong supporters of Croatia's independence began to criticize the anti-democratic aspect of the Tudjman regime. For example, in 1997, the Commission observed that

> the influence of the ruling party (HDZ) goes beyond the political, and is reinforced by state control over the (outrageously) pro-government electronic media, in particular TV. Effective separation of powers, between government, administration and judiciary, does not exist … The delays in the procedure for the admission of Croatia to the Council of Europe (admitted in November 1996) are indicative of (continuing) concerns regarding the state of democratisation and the respect for human and minority rights in Croatia. (European Commission 1997b: 6)

More disturbing for the Brussels administration was the fact that President Tudjman's inaugural address after his re-election in 1997 did not even mention reconciliation, return of refugees and displaced persons and human rights (European Commission 1997b: 7).

Still, it was under the HDZ regime that a Ministry for European Integration was established in 1998 and the *Action Plan for European Integration* published in 1999. Aware that Croatia's progress towards EU membership was going to be conditioned by significant improvements in various areas, the action plan underlined the country's commitment and readiness to cooperate. For example, with regard to the return of refugees, it acknowledged that 'it is essential, to the extent objectively possible, to increase its dynamics, through decisively removing administrative obstacles that still exist, in particular at local levels of government' (Government of the Republic of Croatia 1999: 22). And, when it came to cooperation with the International Criminal Tribunal for the former Yugoslavia (ICTY), the action plan noted that '[t]he Office for the Cooperation with ICTY is established as the expert service … which performs expert and administrative tasks … The Office monitors and co-ordinates and directs the activities of state bodies competent for the co-operation with the Court, and of other legal subjects whose work is of significance in the co-operation with the International Criminal Court' (Government of the Republic of Croatia 1999: 328).

However, in contrast to the phrases offered by the action plan, Croatia's non-compliance generated further international criticism and isolation, a situation that improved only after the change of government in 2000, when Stjepan Mesić, from the Croatian People's Party, was elected president, and Ivica Račan, leader of the Social Democratic Party, became new prime minister. Even though the new government passed a *Declaration on Cooperation with the ICTY*, noting that all war crimes committed in Croatia were liable to the tribunal's jurisdiction and promising full cooperation, that cooperation continued to be characterized by numerous ups and downs. As one author correctly puts it, 'indictments by the ICTY were hugely contentious and divisive among Croatian political elites and society because, for the majority of people living in Croatia, the war was recent, memories were vivid and the indicted war criminals were perceived as respected Croatian war heroes' (Braniff 2011: 108). It was only in 2004 that the EU, after having enthusiastically received the newly elected, reformed, HDZ government and Ivo Sanader, the new prime minister, described Croatia as fully committed to its cooperation with the tribunal.

Croatia signed the EU Stabilisation and Accession Agreement in 2001 and attained candidate status in 2004. According to the Commission's report, the country managed to make significant progress. With regard to the political criteria, although described as a country with 'stable democratic institutions which function properly' and with 'opposition [that] plays a normal part in the operation of the institutions', the report insisted that 'Croatia needs to make substantial improvements in the functioning of the judicial system' and that '[t]he effectiveness of the fight against corruption needs further strengthening' (European Commission 2004c: 37). In addition, as had been pointed out on numerous other occasions, the country had 'to take measures to ensure that the rights of minorities, in particular of the Serb minority, are fully respected' (European Commission 2004c). In contrast to the Tudjman period, the report observed that '[t]he Croatian economy has achieved a considerable degree of macroeconomic stability with low inflation' (European Commission 2004c: 53). The Council welcomed the Commission's opinion about Croatia's application and in addition to the aforementioned aspects, added that 'the achievement of candidate status by Croatia should be an encouragement to the other countries of the Western Balkans to pursue their reforms … The advance of the individual countries of the region towards European integration will proceed in parallel with the regional approach, which remains an essential element of EU policy. The European Council urges Croatia to continue to make a strong contribution to the development of closer regional cooperation' (EU Council 2004: 8).

Following a symbolic delay, caused by the ICTY's March 2005 statement that Croatia was failing to capture one of its war criminals, namely Ante Gotovina, the EU nevertheless opened accession negotiations with Croatia in October, as it primarily wanted to avoid further alienation. The 2006 Commission report on Croatia's progress towards EU membership was generally positive. In fact, in addition to responding to the tribunal's requests, the country progressed in other areas, mainly relating to economic performance. The 2007 November parliamentary elections kept the HDZ in power and, based on the EU's reports, Croatia seemed very open to discuss the Copenhagen political and economic criteria and make the necessary improvements. Thus, at this stage, it was possible to say that the Croatian authorities were becoming more and more committed to the process of EU integration. Accordingly, in November 2008, the Commission proposed an indicative road map that would facilitate further progress in the hope of reaching the final stage of negotiations with Croatia by the end of 2009 (European Commission 2009a: 5). Still, it is important to note that improvements in other areas were lagging behind. For example, as observed, ethnically motivated attacks against the Serb minority 'remain unreported due to a lack of confidence in the authorities and the occurrence of negative sentiment in the wider public towards Serbs. Members of the Serb minority, including those who remained in Croatia during the war, face difficulties concerning access to employment, especially in the war-affected areas' (European Commission 2009a: 15).

In 2011, the Brussels administration announced the target date for Croatia's accession to the EU to be 1 July 2013. By analysing the post-2000 period, it is possible to say that the country made impressive progress. At one point, having realized the benefits of the whole process and eventual EU membership, the main political figures and their respective parties in Croatia started to pay much more attention to what remained to be done in order to join the EU as soon as possible rather than debating whether they should join it at all. In addition, as correctly observed in the literature, the ever-growing interest since 2000 in joining the EU has been the result of a general understanding that only EU membership will represent 'the final stage in the process of making Croatia an independent and sovereign state' – a status that has often been challenged by the ICTY's interference and the EU's conditionality policy (Jović 2011: 40). In a referendum in January 2012, the Croatian voters clearly supported their country's entry into the EU. As *Večernji list*, the main Croatian conservative newspaper, put it, the country was choosing between Europe and the Balkans, two extremely different paths. More precisely, many voters were choosing between political stability and economic progress, on the one hand, and stagnation, often associated with the Yugoslav wars and, more importantly, victimhood, on the other.

Macedonia

The first free elections in Macedonia were held in November 1990 and were followed by immediate concerns as the dominant nationalist party, Vnatrešna makedonska revolucionerna organizacija – Demokratska partija za makedonsko nacionalno edinstvo (VMRO-DPMNE), did not obtain the clear majority necessary to form a government. However, after three rounds of voting and obvious parliamentary deadlock, some sort of compromise was achieved and a government of experts formed, with Nikola Kljusev as prime minister and Kiro Gligorov as president. Following a referendum on secession in September 1991, Macedonia declared independence in December.

The process of international recognition of the new Macedonian state – severely affected by numerous Greek protests generated by the country's new name and consequent embargo, represented the main challenge of the new leadership. As one study summarized the initial steps:

> [m]ost observers agreed that the EC states supported Greece on the Macedonian issue, not because of the merits of the Greek case, but because the EC Council of Ministers recognized the right of member states to exercise an unofficial veto on issues that affect their national interests. Furthermore, in exchange for EC support on the Macedonian issue Greece promised to ratify the Maastricht Treaty … In addition, Greece also promised to support EC sanctions against Serbia, a traditional ally of Greece, and to ratify an EC financial protocol with Turkey, a traditional enemy of Greece. (Danforth 1995: 150)

Nevertheless, soon after the official recognition of Slovenia and Croatia in January 1992, the Brussels administration expressed its interest in cooperation with the authorities of Skopje 'capable of improving political stability and economic progress in the area' (EPC 1992c), so much needed to avoid 'the unintended consequences of UN sanctions' (EPC 1992e). Consequently, the Europeans agreed to grant substantial economic aid to Macedonia, and in January 1993, Gligorov met EU officials to discuss the political situation in the region, implementation of Community aid and trade between Skopje and Brussels (European Commission 1993: 265).

In Macedonia, this was a period clearly pointing to the country's fragility. In summer 1992, a no-confidence vote led to the resignation of Kljusev's government, mainly because of its lack of progress in securing recognition. A new coalition government, led by Branko Crvenkovski, the head of the Social Democratic Union of Macedonia, decided to work more closely with the Brussels and Washington administrations, as well as domestic political factions, hoping to find a solution to the home-grown antagonisms, but without any significant success. In fact, aware of the problems, President Gligorov managed to convince the UN first to deploy a preventive mission, UNPREDEP (initially UNPROFOR), and then to extend its mandate, primarily to monitor borders and prevent internal clashes (UN 1996). For example, the ethnic Albanians insisted on having their university curriculum in Albanian, a project that started to materialize with the establishment of their own university in Tetovo in 1994 – a non-constitutional move. The incidents surrounding the opening of the university generated other types of confrontation across the country. In fact, as concluded elsewhere, '[t]he clash intensified the polarisation of inter-ethnic relations in the country and increased concerns the radical Albanian leadership would succeed in "homogenising" the ethnonationalist agenda of the Albanian community and integrating its various political parties' (Leatherman et al. 1999: 162).

Although the 1994 elections, characterized by boycott, fraud, internal splits and public scepticism, kept Crvenkovski's government in power, numerous corruption scandals that followed affected its popularity dramatically. As one author puts it: 'Even its Western supporters could not rescue it now, but to the extent that they did try to save a political configuration that was perceived as corrupt and ineffective in the eyes of the public, it delegitimized the West and the external project of liberalism itself. So it was no surprise when in the 1998 elections, over 45 percent of the vote went to a coalition of VMRO and a newly formed party, the Democratic Alternative' (Boduszyński 2010: 149).

When talking about the European Union position, one report assessed the situation in Macedonia in the period 1996–2001 and concluded that '[t]he future is highly uncertain – if open conflict can be avoided, political reform will be necessary to address ethnic tensions and further economic reform' (Investment Development Consultancy/Development Strategies 2001: 1). More precisely, with regard to political developments, Macedonia was portrayed as a country where '[p]olitical competition is still seen very much as "winner-takes-all"; parties seek to gain control of state institutions for the sake of the economic rents they provide. As a consequence, the state remains highly centralised and highly politicised. Meanwhile, civil society is weak and provides little in the way of checks and balances on government conduct' (Investment Development Consultancy/ Development Strategies 2001: 2). With regard to ethnic tensions directly affected by a political and governance problem, Macedonia was described as 'a deeply divided society' in which 'the rural poor – and ethnic Albanians in particular – have little stake in FYR Macedonia as a nation' (Investment Development Consultancy/Development Strategies 2001: 3). Finally, Macedonia's economic performance was 'extremely weak, with high unemployment and a large number of inefficient state-owned enterprises' (Investment Development Consultancy/ Development Strategies 2001). In addition to all these issues, the report assessed the EU's approach towards Macedonia as being without a clear strategy and rather risky: 'Little attention was given to central issues such as ethnic tension and the dysfunctional nature of the political system. In addition, no country strategy was prepared in the year 2000. The recent shift in emphasis towards EU integration objectives has not been properly analysed and has led to even more complex and disperse programmes, which the EC is unlikely to be able to implement successfully' (Investment Development Consultancy/Development Strategies 2001: 7). Thus, European involvement should have primarily focused on Macedonia's domestic issues and thus supported political reform, ethnic reconciliation and economic development, 'leading in turn to a socially cohesive and economically dynamic Macedonia that will be in a position to make a success of the SAA and eventually of integration in the EU' (Investment Development Consultancy/Development Strategies 2001: 12).

When signing the SAA in April 2001, the Brussels administration outlined priority areas to be addressed, underlining the relevance of making the population aware of the benefits of the implementation of the SAA: 'Greater efforts need to be made to explain that EU integration is based on democratic principles, including respect for minority rights, hence the involvement of the EU in trying to resolve the current inter-ethnic crisis in the country' (European Commission 2002a: 31). Indeed, in 2001, it became even more obvious that Macedonia was a country characterized by uncertainty, given the armed insurgence when the Albanian rebels organized into the National Liberation Army started insisting on more rights. As rightly warned by the EU, 'the country was several times near to civil war' (European Commission 2002a: 4). Apart from the name issue, which was still present but often ignored because the provisional name temporarily appeased the interested parties, the Macedonian authorities had to address their Albanian minority.

After the signature of the Western-sponsored Ohrid Agreement between the two parties in August 2001 and the almost immediate deployment of NATO peacekeepers, the 1991 constitution, which had originally presented Macedonia as the state of the Macedonian people, was amended. It improved the status of the Albanians, including their language, representation in the public sphere, educational rights and so on (Job 2002: 89–93). The EU expressed its open appreciation for the Macedonian government for having managed to address some rather sensitive issues and for having repeatedly shown commitment to the EU, which in turn helped the Europeans to be perceived more favourable by the public. The following statements were even more optimistic as the closer relations between the two parties were described as 'a guarantee of economic development and further political stabilisation' (European Commission 2004a: 36).

Macedonia gained candidate status for EU membership in 2005. In its first report on progress made, the Commission, amongst many, applauded Macedonian authorities for new constitutional amendments as they improved the situation in the judiciary and the country's commitment towards the ICTY. However, there were still numerous areas that required significant improvements. Apart from the elections that were again characterized by irregularities, causing sharp tensions between the majority (led by VMRO-DPMNE) and the opposition (mainly DUI-PDP), a number of other observations were very critical of Macedonia's ability to proceed with the obligations of EU membership, such as the fight against corruption, improvement of social policy and employment, promotion of competition policy, and so on (European Commission 2006).

Documents discussing Macedonia's commitment and progress towards EU membership have always clearly identified both the areas that were performing well and the ones that necessitated more reform. However, what they often reveal is the existence of interruptions and thus inconsistencies. For example, in one of the reports, the Commission observed the following:

> Objective and merit-based criteria are not consistently used in recruitment and promotion. There is a lack of transparency and accountability in recruitment decisions by individual administrative bodies and the CSA [Civil Servants Agency] lacks mechanisms for assuring the legality and regularity of those decisions. Senior management positions are often filled externally without sufficient regard to professional qualifications and experience, in contravention with the law on the civil service. (European Commission 2008b: 11)

Even though the consequent report already talked about some improvements and apparent merit-based recruitment and equitable representation, overseen by the CSA, the one released in 2010 was not optimistic: 'The capacity of the CSA to take on the new competences conferred by the Law on public servants is insufficient in terms of expertise, budget and available infrastructure' (European Commission 2010c: 10).

A number of inconsistent observations clearly confirms Macedonia's own struggle to progress continuously to the extent of being offered full EU membership. By looking at various reports, including the latest ones, it is possible to conclude that Macedonia can claim limited progress, meaning that its accession and integration in the EU is likely to be a rather lengthy process. According to one study, the current Macedonian government bears enormous responsibility for the country's slow progress: 'Four years after VMRO-DPMNE came to power [2006], there were unprecedented and widespread conflicts of interest in Macedonia's transitional society and what seemed to be a real plague of systematic corruption, favouritism in public spending and cronyism of the worst kind' (Ordanoski 2011: 97).

In contrast to Slovenia and Croatia, two economically-superior republics of the Yugoslav federation, the situation in Macedonia was and has remained far more complicated. More importantly, '[t]he government had planned nothing that could create economic added value in the future or start a new economic cycle. Many ceremonial ribbons were nevertheless cut and dozens of ceremonial holes dug – empty holes that remain scattered around the country in testimony to pre-election promises' (Ordanoski 2011: 102). In addition, while the Slovenian and, to a lesser extent, the Croatian authorities had a very clear European perspective (inspired by further progress and detachment from the Balkans), the Macedonian leadership seems to be less clear about its objectives. In particular, too much emphasis is placed on domestic policy and not enough on foreign policy and potential benefits. Although it is true that neither Slovenia nor Croatia was involved in any post-Yugoslav name issues, recent inter-ethnic violence or alarming unemployment rates, Macedonia should still recognize that it is thanks to the EU that the country has managed to progress and become more legitimate than in the early 1990s and thus it should continue to invest even more effort in its European future.

Montenegro

During the Yugoslav state crisis, when the Serbian authorities were continuously depicted as the main aggressors, the position of Montenegro was not an easy one. It shared much of Serbia's destiny, first as a founding member of the Federal Republic of Yugoslavia (FRY), until 2003, and then as a constituent party of the State Union of Serbia and Montenegro. However, even as early as the late 1990s, the first political discords between the Montenegrins and Serbs were already beginning to dominate the dialogue between them and question future cohabitation. The then Prime Minister and soon after President of Montenegro, Milo Djukanović, played a crucial role in the process. He abandoned his original support for the Serbian leadership and developed a platform for redefinition of relations within the FRY.

Although the platform called for significant adjustments in relation to the governing responsibilities, it still saw Montenegro together with Serbia. While the majority of citizens of Serbia seemed to be in favour of the content of the platform, the Serbian leadership labelled it as unconstitutional. In an interview given to the Serbian weekly *Vreme* a couple of months after the termination of NATO bombing of the FRY, Svetozar Marović, President of the Montenegrin Parliament, criticized the Serbian authorities for being extremely self-centred and thus without any intent to address their fellow citizens and to converse with the Montenegrins about their position and expectations. Alongside these observations, he clearly stated that if Serbia was not going to respond positively to the platform, Montenegro was ready to consider a referendum on independence (Marović 1999).

The period following the 1999 Kosovo crisis was characterized by various political divisions of Montenegro (confirmed by the local elections) as well as by growing uncertainty with regard to its relations with Serbia (confirmed by the federal elections). However, when Slobodan Milošević was removed from power in Serbia in October 2000, the Montenegrin leadership openly supported new democratic forces. Still, as observed in the literature,

> [d]espite the collective relief and easing of tensions, there was an understanding that Montenegro's strategic significance had been reduced. More crucially, it removed a key component of the [Djukanović-led] DPS's [Democratic Party of Socialists] justification for control over all spheres of society – the threat from Belgrade … had diminished. Post-October 2000, Montenegrin leaders knew that if they were to continue along the road of independence, they could no longer rely on the unquestioning support of the USA and EU. (Morrison 2009: 181)

Concern surrounding the status of Montenegro was postponed by the EU-mediated Belgrade Agreement, in March 2002, and the establishment of the State Union of Serbia and Montenegro, in February 2003. For the Brussels administration, such an arrangement represented the right way to maintain regional stability. However, 'given that the formation of the new joint state was initiated and implemented by political elites in Serbia, Montenegro and the EU and not a democratic referendum, many people in Montenegro regarded the creation of SCG [Srbija i Crna Gora] as an elite-driven imposition that did not reflect the will of the people' (Morrison 2009: 188). For Djukanović, the internal frustration and uncertainty caused by the Belgrade Agreement meant looking for an alternative strategy: 'The Belgrade Agreement raised Montenegro's potential to act as a "nuisance" to Serbia, which could have induced the latter to secede. This would have left the international community with little choice but to bless the velvet divorce between the two republics' (Tocci 2007: 87).

In May 2006, a referendum on independence confirmed Montenegro's ambition to continue on its own. The pro-independence movement, led by Branko Lukovac from the Democratic Party of Socialists (DPS) and including other political parties (Social Democratic Party, Democratic Union of Albanians, Democratic League in Montenegro, Civic Party and Liberal Party), maintained that the behaviour of Serbia's political elite and its unenviable international reputation was capable of affecting Montenegro's progress negatively. Following the attainment of independence, both the government and parliament of the new state presented the process of European and Euro-Atlantic integration as their top priority.

The Brussels administration, aware of its crucial role in the whole process, expressed appreciation for Montenegro's decision and welcomed the adoption of the new constitution. Apart from noting that '[t]he referendum was conducted in line with international standards', the EU's official report, more importantly for the new Balkan state, stressed that numerous areas necessitated improvements; for example, 'the lack of confidence in the system and concern over the level of political influence exercised over the judiciary are still high' and 'the situation concerning the fight against corruption calls for urgent action in order to achieve results on the ground' (European Commission 2006: 6, 10–11). Thus, it was made clear that Montenegro's path towards EU membership would clearly be conditioned by various fulfilments. Following the signature of the SAA, the first step in the process enabling Montenegro to receive significant pre-accession financial support (under the Instrument for Pre-Accession Assistance), in October 2007, Olli Rehn, the then EU Commissioner for Enlargement, underlined that 'growth depends on investments, and investors need legal and political security, transparency and predictability. Key for this would be strengthening of the rule of law and an efficient fight against corruption and organized crime … EU assistance will hasten the development of better institutions and administrative capacities' (Rehn cited in Radulović 2011: 81–2).

The 2008 progress report acknowledged the efforts of the Parliament of Montenegro to establish a Council for EU integration to monitor the implementation of the SAA and European integration process. However, according to the report, 'parliament has not made good use of its working bodies and observance of the rules of procedure remains problematic. Furthermore, inspection and control mechanisms provided in the rules of procedure are not being used appropriately. Parliamentary work is not yet efficient and transparent enough' (European Commission 2008c: 7). Nevertheless, Montenegro applied for EU membership in late 2008. The subsequent report noted that some progress had been made, but this was far from being enough. As observed, 'corruption remains prevalent in many areas and continues to be a particularly serious problem. The commitment of the authorities has not yet been backed by consistent implementation of anti-corruption legislation' (European Commission 2009b: 13).

In November 2010, the European Commission's opinion on Montenegro's application for EU membership was generally positive. While noting that 'Montenegro's accession would have a limited overall impact on European Union policies and would not affect the Union's capacity to maintain and deepen its own development', the Commission recommended the opening of negotiations 'once the country has achieved the necessary degree of compliance with the membership criteria' (European Commission 2010d: 12). Still, it outlined seven priority areas requiring significant progress: the legislative framework for elections, public administration reform, strengthening of the rule of law, the anti-corruption measures, strengthening of the fight against organized crime, enhancement of media freedom and the implementation of anti-discrimination policies (European Commission 2010d). These areas, however, did not prevent EU leaders from granting candidate status to Montenegro in December – a decision confirming the Union's readiness to assist aspiring EU candidates and members.

In order to demonstrate its full commitment to progress and EU membership, the Montenegrin authorities adopted an *Action Plan for Monitoring Implementation of Recommendations Given in the European Commission's Opinion* in February 2011, with the aim of addressing all the priorities before the new EU evaluation – an extremely ambitious objective given 'the relatively weak position of the new government with regard to introduction of reforms in the judiciary and the fight against organised crime and high-level corruption' (Radulović 2011: 92). In their new report, the Europeans assessed the implementation of the action plan as consistent and systematic, but still noted that '[t]here continues to be a need for stronger monitoring of corruption and conflicts of interest in the judiciary. Corruption remains prevalent in many areas and continues to be a serious problem' (European Commission 2011a: 14–15).

The EU decided to open accession negotiations with Montenegro in June 2012. Although it is difficult to say how long it will take for the smallest in population of all post-Yugoslav states to become a full EU member, it is possible to argue that the Montenegrin case has managed to advance significantly. The country, at least, and in contrast to some other potential EU candidates and members in the Balkans, has had a clearer EU vision. Such a vision served as a backbone in the process of attainment of independence and, more importantly, has remained since then and thus has helped convince the Brussels administration of Montenegro's aspirations for European integration.

Serbia

Throughout the 1990s, Serbia and, more importantly, the Serbs, were continuously portrayed as the vampire nation with their 'vampire-in-chief', Slobodan Milošević (Longinović 2011). However, the overthrow of Milošević in October 2000 provided the Serbs with an opportunity to change their well-established image abroad. Consequently, the EU enthusiastically decided to reward the newly elected anti-Milošević Democratic Opposition of Serbia (DOS) by lifting its economic sanctions, securing reconstruction assistance, providing aid packages and signing trade agreements. The new Prime Minister, Zoran Djindjić (whose main success was the arrest and transfer of Milošević to the ICTY), kept saying that there was no time to waste. In his view, Kosovo was de facto independent and Serbia had to move on with the processes of democratization and Europeanization. By contrast, the new President, Vojislav Koštunica, who was also welcomed by the European officials as a symbol of new democratic orientation, did not fully agree with Djindjić's approach. For example, he rejected the urgency of the Kosovo status, claiming that it was better to 'wait for another five years, because the later this issue is addressed the better it will be for [the Serbs]' (UNMIK 2003). In addition, the delayed cooperation with the Hague tribunal was confirmation of Koštunica's reluctance to break with the Milošević era: on various occasions, he accused the tribunal of undue interference and assault on the dignity of his state (Rupnik 2004: 105). Moreover, Koštunica's rhetoric influenced the Serbian public to the extent that the majority objected to any cooperation with the ICTY.

The assassination of Prime Minister Djindjić in 2003 represented an enormous loss for Serbia: the processes of democratization and Europeanization were immediately interrupted. The European officials were naïve enough to believe that Koštunica, the new prime minister, was ready to make a big step towards European integration, shifting from introverted post-communist conservatism to a modern, democratic and open society. Indeed, they continued to express their support for Serbia, leading to the opening of negotiations for the SAA in October 2005. However, given an obvious lack of cooperation with the Hague tribunal, the negotiations were suspended in May 2006 and resumed only after the victory of President Boris Tadić's Democratic Party in the 2007 Serbian parliamentary elections.

In this period, Serbia's compliance with the EU's policy of conditionality was often assessed as insufficient. However, based on the number of official statements, it was not the Europeans who lacked a clear vision with regard to the future of Serbia, but the Serbs themselves. This was primarily due to the diametrically opposing views among domestic political elites. For example, Vojislav Šešelj's Serbian Radical Party and, to a lesser extent, Vojislav Koštunica's Democratic Party of Serbia and numerous former members of Milošević's Socialist Party of Serbia often encouraged anti-European feelings. In their view, the process of Europeanization and eventual EU membership was not something Serbia should pursue. In order to justify such a standpoint, their statements were often inspired by European (and American, of course) involvement in the Yugoslav state crisis of the early 1990s and, more importantly, in the 1999 NATO bombing of Serbia. On the other hand, Tadić's Democratic Party and Tomislav Nikolić's Serbian Progressive Party have generally been pro-European, although some of their members' statements and actions have occasionally undermined their apparent commitment to the process of Europeanization.

Given that the elites do not share similar standpoints with regard to Serbia's European perspective, the process of Europeanization is further complicated. One study, while pointing out that in transitional systems 'consensus about basic social, political, and economic priorities and values is often absent', sees public opinion surveys as a useful way to discover what society really thinks and what its main concerns are (Farkas 2007: 40). For example, back in 2004, the Serbian

Government Office for European Union Integration conducted a study in order to show how the Serbian public approached Europe. Respondents were divided in four categories: the Euro-enthusiasts, who said that Serbia must make every effort to join the EU; Euro-realists, who perceived integration in the EU as a necessity; Euro-sceptics, who questioned the intentions of Europe and the West in general; and Europhobes, who feared the dominance of Western Europe and thus strongly opposed integration. Based on the results, Euro-realists were the dominant category, while Europhobes were the least popular (Government of the Republic of Serbia 2004). Later, following Kosovo's declaration of independence in February 2008 and the immediate decision of many EU members to recognize it, there were many in Serbia that clearly questioned European policy towards the region. In fact, some studies showed that support for joining the EU continued to decrease: for example, in December 2010, 57 per cent of respondents were in favour of Serbia's EU membership, whereas in December 2011, this number had dropped to 51 per cent (Government of the Republic of Serbia 2012).

Soon after Kosovo's proclamation of independence, accompanied by demonstrations and attacks on foreign diplomatic missions in Belgrade, acts seriously condemned by the Brussels administration, the EU signed the SAA with Serbia in April. The Europeans reproduced their earlier rhetoric presented by the Council, stating that 'a stable and prosperous Serbia fully integrated into the family of European nations is important for the stability of the region' (EU Council 2008a: 20). Back then, it was obvious that Serbia could make some further progress, however, many areas were intentionally ignored due to the high level of corruption. As assessed, 'corruption remains prevalent in many areas and continues to be a serious problem. There are major obstacles to rooting out systemic corruption. The most vulnerable sectors are public procurement, privatisation and other large budgetary expenditures, but also taxation, customs and licensing' (European Commission 2009d: 12). In addition, and in contrast to some other post-Yugoslav states, Serbia was expected to intensify its cooperation with the Hague tribunal, given the tribunal's capacity to affect Serbia's EU path: 'The need for an increase in Serbia's operational capacities and the adoption of a more rigorous and multi-disciplinary approach to arresting the fugitives was also highlighted by the ICTY report. Full cooperation with the ICTY remains an essential condition for membership of the EU' (European Commission 2010b: 18).

As one study correctly assessed, since the breakup of the Yugoslav federation, 'Serbia has been invariably late: late in recognizing the spirit of change in 1989, late in reacting to Milošević's devastating policies, late in seeing the reality in Kosovo, late in accepting the cooperation with The Hague as a conditional *sine qua non*, late in defining the EU integration as the highest priority and hence late in conducting absolutely necessary reforms' (Ristić 2009: 122). Indeed, various indexes and indicators, such as the 2010 Democracy Index, the Global Competitiveness Index and the Worldwide Governance Indicators, have demonstrated that Serbia has been late. The only significant progress can be associated with electoral democracy and elections. Still, elections cannot consolidate democracy; they only serve to confirm democratic legitimacy. What is needed is a strong link between democracy (free and fair elections) and constitutional liberalism (rule of law and limited power) that would lead to the establishment of permanent institutions characterized by a depoliticized public sector and independent courts (Vujačić 2011: 22). Here, although European Union involvement can provide assistance and apply its policy of conditionality, it is the Serbian leadership that decides on the level and speed of cooperation with both the EU and the Hague tribunal and, accordingly, the more they are ready to cooperate, the bigger the rewards will be. While cooperation with the Hague tribunal has progressed satisfactorily (Radovan Karadžić was arrested and extradited in 2008, and Ratko Mladić and Goran Hadžić were sent to The Hague in 2011), resulting in Serbia's EU candidacy in 2012, the Kosovo issue still requires great attention.

Even though the European Commission recommended that Serbia should be granted EU candidate status, it also outlined the role of the EU in relations between Serbia and Kosovo: 'The EU will continue to facilitate the dialogue between Belgrade and Priština to promote cooperation, achieve progress on the path to the EU and improve the lives of people. The dialogue has proved to date a valuable means to address differences between the two parties and has led to several agreements in areas including freedom of movement of goods and persons, civil registry and cadastre' (European Commission 2011c: 27). Such a disputable statement was further eroded by the EU's assessment that 'the integration of Kosovo Serbs has improved in the South, [while] tensions in northern Kosovo have increased' and suggestion that '[p]eople in the north also need to benefit from the European perspective' (European Commission 2011c: 28). What the EU has actually tried to suggest is that the Serbs inhabiting northern Kosovo should accept Kosovo's independence. With this in mind, the Commission expressed its willingness to support 'Kosovo's efforts to further fulfil its European perspective' and 'the development of relations between the EU and Kosovo caused by differences on Kosovo's status' (European Commission 2011c: 28).

The relationship between the EU and Serbia, via Kosovo, has been rather difficult and is likely to remain so. The Serbian authorities have refused to recognize Kosovo's independence, although their standpoint could cost them membership of the EU. The Serbian position has also been encouraged by the fact that five EU members (Cyprus, Greece, Romania, Slovakia and Spain) have not recognized Kosovo either, mainly because of their own domestic issues. However, other EU representatives seem to be much more concerned about how to change the position of those five rather than think about possible scenarios such a change might generate.

Bosnia and Herzegovina

The Europeans deluded themselves that the independence of Bosnia and Herzegovina in April 1992 would calm down the ultranationalists, bring about an acceptable compromise and, more importantly, prevent large-scale violence.

Their decision, accompanied by various statements and declarations, called for an immediate ceasefire. For example, only a few days after the recognition, they 'call[ed] upon Serbian and Croatian Governments to exercise all their undoubted influence to end the interference in the affairs of an independent republic and to condemn publicly and unreservedly the use of force in Bosnia and Herzegovina' (EPC 1992a), warning them that '[p]arties responsible for such actions will be internationally held accountable for their acts' (EPC 1992b).

The proposed and negotiated attempts – the 1992 Carrington–Cutileiro peace plan, the 1993 Vance–Owen and Owen–Stoltenberg plans and the 1994 Contact Group plan – failed to bring the warring parties to a resolution and prevent the July 1995 Srebrenica massacre, the worst genocidal atrocity in post-Second World War Europe, which left the image of the Serbs as aggressors and the Muslims as victims. It was later in December that the US-brokered Dayton Peace Accords put an end to the war in Bosnia and Herzegovina by dividing the country into two semi-independent entities: the Federation of Bosnia and Herzegovina, inhabited primarily by Bosnian Muslims and Croats, and the Serb-dominated Republika Srpska (Serb Republic).

The situation in post-Dayton Bosnia has dominated numerous debates. Current opinions tend to agree that the accords, although successful in providing an immediate solution to the crisis, somehow ignored the possible consequences:

> Post-Dayton BiH [Bosnia and Herzegovina] is a post-modernist witticism from the geopolitical genie that burst out of the bottle in Dayton, a tentative, stop-gap state in the guise of a sovereign, self-sustaining state. What this means is that the greater part of the Dayton Peace Agreement is not a new state of Bosnia and Herzegovina as the creation of the Agreement's architects, but rather a creative interpretation of the work (BiH) and its potential, in the quantitative sense, far in excess of the rendition itself; it is this, at the end of the day, that has given rise to the crisis of the work, the crisis of BiH, for no interpretation can substitute for something that does not really exist. (Ćurak and Turčalo 2012: 63–4)

The Dayton Agreement did not establish an executive authority responsible for policy making and problem solving, but rather it contributed to the already deep divisions within Bosnian society. Still, the establishment of the Office of the High Representative, an ad hoc international body supervising implementation of civilian aspects of the agreement, and awareness of the overall complexity of the Bosnian case, made the EU think about a more systematic approach focused on political stability and economic recovery. In 1997, following the establishment of political and economic conditions to be fulfilled by the countries of South Eastern Europe, the EU's report on the compliance of Bosnia and Herzegovina observed that since the Dayton Agreement, 'it has not proved easy to persuade the leaders of any of the three communities to commit themselves wholeheartedly to democratic co-existence with the others' (European Commission 1997b: 3). Lack of such a commitment affected the implementation of legislation regarding human rights, rule of law and refugee return: 'International organisations report numerous cases of discrimination against existing members of ethnic minorities by public authorities and employers. These include discrimination between the two ethnic groups within the Federation ... Moves by the RS authorities to insist on use of the Cyrillic alphabet constitute a further handicap for non-Serbs' (European Commission 1997b: 4).

The EU launched the Stabilisation and Association Process with Bosnia and Herzegovina in late 2000 and thus gave the local authorities the opportunity to strengthen the central level of government in order to come closer to the EU. Since then, the situation and potential of the country have been greatly discussed. As some authors put it, although the country has managed to overcome its most difficult peace-building challenges, 'the international mission in BiH has developed into a quasi-protectorate, with extensive international powers and some strikingly illiberal tendencies' leaving the country 'isolated from the European integration processes taking place around it' (Knaus and Cox 2004: 55). The external involvement created a state without its own strong political leadership and functioning institutions, leaving an impression that the country will continue to stagnate and is unlikely to become an effective state. Even the 2003 Stabilisation and Association Report acknowledged the EU's exaggerated presence: 'EU political leaders figure prominently in Bosnian public life (along with other international actors). Admittedly, however, the multiplicity of European or EU-related structures ... probably lead to confusion in the minds of both general public and the elite' (European Commission 2003: 33). Still, both this and the following report presented the EU's large-scale presence as necessary and beneficial. This went so far as to claim that, even though there were no relevant opinion polls on EU issues conducted, 'the public perception of the EU is generally positive' and that '[r]eforms have also highlighted the EU role in the public mind' (European Commission 2004b: 16–17).

Later, the Europeans made an assessment that Bosnia and Herzegovina was managing to make only limited progress towards meeting European standards. For example, apart from 'mak[ing] no progress towards creating more functional and affordable State structures which support the process of European integration', the country's state-level government structures were 'unable to overcome internal conflicts – often based on ethnic or Entity allegiance – and to avoid deadlocks' (European Commission 2007: 8, 10). On the other hand, other, non-EU-drafted analyses criticized the EU's involvement for having 'pragmatically appealed to the common sense of domestic political leaders, trying to make them see the virtues of undertaking the reforms themselves rather than presenting the reforms as a sacrifice they have to accept in order to be admitted to the club' (Noutcheva 2007: 7). Also, with regard to the democratization and Europeanization processes, there should have been much more emphasis on the people and their everyday concerns: 'In order for democracy to become "the only show in town" it is inevitable that the principles of democracy find their way directly into the people's attitudes towards their governments and that they are carried by pro-democracy actions and deeds of elites and ordinary citizens alike' (Džihić 2007: 27).

The EU signed the SAA with Bosnia and Herzegovina in June 2008. As summarized by the Council, the agreement was aimed at democracy and the rule of law, political, economic and institutional stability, political dialogue between the two parties, international cooperation in the country, regional cooperation, and so on (EU Council 2008b). However, the complex institutional and political climate in the country continued to cause delays. This was also due to the corruption that was present in numerous areas, 'especially within government and other State and Entity structures, linked to public procurement, business licensing, in the health, energy, transportation infrastructure and education sectors' (European Commission 2009c: 14–15). That the progress was slow could be concluded by reading later reports: in Bosnia and Herzegovina, although considered 'a potential candidate for EU membership', '[t]he overall pace of reforms has been very limited' and '[t]he political representatives lack a shared vision on the direction to be taken by the country' (European Commission 2011b: 3, 4, 10).

As discussed in the literature, the Europeanization approach, assuming that the Bosnian authorities would fully commit themselves to the necessary transformation, inspired by potential membership of the EU, has failed to materialize. According to one study, we can talk about three main challenges that clearly confirm the failure of the EU's approach towards Bosnia and Herzegovina (Tzifakis 2012). First, the Europeans have overlooked the difficulties surrounding the Europeanization process of an ethnically-divided state in which all three ethnic communities, although supportive of the country's EU path, are not ready to make compromises that could potentially affect their own position – a situation that is further complicated by the impression that the Brussels administration has been biased towards the Bosnian Muslims and very critical of the Bosnian Serbs (Tzifakis 2012: 137–8). The second challenge relates to the credibility of the accession conditionality as '[t]he Bosnians do not believe that they have a credible short-term membership perspective; the EU has repeatedly failed to stand behind a single uncontested position; and, more worryingly, Brussels has demonstrated that it is ready to reward even partial or limited compliance' (Tzifakis 2012: 139). Thus, the EU's lack of a clear standpoint has empowered the Bosnian authorities to select the areas in which they want to comply and the degree of compliance. Finally, although present through various missions in the country, the EU has failed to deliver clear messages on how to meet EU standards. As noted, '[t]he EU seems to have acted under the assumption that Bosnia's socialisation is a self-generated unproblematic process. However, this has not been the case and the EU shares some responsibility for the fact that the Bosnians have learnt "how not to comply" (rather than "how to comply") with EU rules' (Tzifakis 2012: 141).

In contrast to other post-Yugoslav EU-aspiring states, Bosnia and Herzegovina and Kosovo (next section) have faced 'a "triple transition" (from war to peace, from humanitarian aid to sustainable development and from a socialist political systems and centrally planned economies to democracy, civil society and a free-market economy), which makes it extraordinary difficult for both international and local actors to efficiently address the multiplicity of interrelated and complex challenges' (Džihić 2007: 24). With this in mind, it is possible to argue that Bosnia and Herzegovina will have to invest much more effort in its journey to EU membership than some other post-Yugoslav states. As for EU involvement in the country, as noted in the 2011 Progress Report, 'the Foreign Affairs Council underlined the fact that the EU Police Mission (EUPM) and EUFOR Althea are important elements of the EU overall strategy for Bosnia and Herzegovina' (European Commission 2011b: 5). While it is clear that European presence provides training and helps the fight against organized crime and corruption, and thus supports the local passivity, it is difficult to predict what the consequences of the abolishment of such an arrangement would look like.

Kosovo

It was before the outbreak of the Yugoslav state crisis that full Serbian control over the province of Kosovo was re-established. The takeover of institutions provided the Kosovo Albanians with a rationale for the creation of their own parallel state, resulting in the proclamation of Kosovo first as a republic within Yugoslavia, in 1990, and then as an independent state, in 1991. While observing and getting involved in other parts of the Yugoslav federation, the Europeans also became aware of the problematic situation in Kosovo. For example, in summer 1992, one of the declarations on the deteriorating situation across the Yugoslav federation addressed Serbia's southern province, as well:

> The Community and its Member States noted that the situation in Kosovo is potentially dangerous and urge all parties to show the necessary restraint and sense of responsibility. They urge the authorities in Belgrade to refrain from further repression and engage in serious dialogue with representatives of Kosovo. Failure to do so would impede their prospect for the restoration of normal relations with the international community. The Community and its Member States recall that frontiers can only be changed by peaceful means and remind the inhabitants of Kosovo that their legitimate quest for autonomy should be dealt with in the framework of the EC Peace Conference. (EPC 1992d)

Later, deeper ethnic antagonisms led to the conflict outbreak in Kosovo in 1998, culminating in January 1999 when the Serbian military forces committed a crime against humanity by killing 45 civilians in Račak. Understandably, there was a clear transatlantic ambition to prevent President Slobodan Milošević from directing another episode of ethnic cleansing and achieving full control of the province. Having experienced Slovenia, Croatia and especially Bosnia and Herzegovina, the Westerners claimed to be better prepared to deal with the Kosovo issue. According to Joschka Fischer, the then German Foreign Minister, acting politely with Belgrade officials would lead only to more mass graves, so he stated that the use of force should be considered: 'I am not a friend of using force, but sometimes it is a necessary means of last resort. So I am ready to use it if there is no other way. If people are being massacred, you cannot mutter about having no mandate. You must act' (Fischer cited in Daalder and O'Hanlon 2000: 75).

The consequent NATO military intervention was subject to numerous criticisms. Many commentators described the attack as an aggression against a sovereign state that had not attacked another sovereign state, as, for example, Iraq did when it invaded Kuwait in 1990. In addition, there were many other countries around the world that had been involved in or contributed to similar or even worse atrocities than Serbia was accused of, and in some nations such violations were still occurring, but most were largely, perhaps hypocritically, ignored compared to the Kosovo crisis, although they also presented a strong case for humanitarian intervention (Shah 1999). With this in mind, NATO's intervention in Kosovo had nothing to do with humanitarian impulses and was all about defending the West's geopolitical interests in the region (Chomsky 1999). Other commentators went even further and perceived the aggression as a war of expansion by NATO, a war designed to push United States power right up to the borders of Russia (Gowan 2000). Thus, the intervention was criticized as a colossal error, an example of a policy applied too late, in the wrong place, and in ignorance of history. It was inconsistent and perceived as something that would create problems regardless of whether the outcome was a failure or success (Hammel 1999).

Once the intervention had terminated, the European Union saw the region impoverished and in need of aid. It persuaded the international community to collaborate in the Stability Pact for South Eastern Europe, which was adopted in Cologne in June 1999. Back then, Javier Solana, the EU's High Representative for Common Foreign and Security Policy, reminded Kosovo Albanians that independence was not on the agenda and that technically Kosovo was still part of Yugoslavia. He argued that the main task of the international presence was to establish standards first, and then to discuss the final status of the province.

The United Nations Security Council Resolution 1244, passed on 10 June 1999, established the UN Interim Administration Mission (UNMIK), exercising full executive, legislative and judicial role. The resolution declared the

> establishment of an interim administration for Kosovo as a part of the international civil presence under which the people of Kosovo can enjoy substantial autonomy within the Federal Republic of Yugoslavia, to be decided by the Security Council of the United Nations. The interim administration was to provide transitional administration while establishing and overseeing the development of provisional democratic self-governing institutions to ensure conditions for a peaceful and normal life for all inhabitants in Kosovo. (UN 1999)

However, neither UNMIK nor the NATO-led peacekeeping Kosovo Force (KFOR) was capable of preventing violations of human rights in the areas comprising the remaining Serbs and other non-Albanian population. Thus, in contrast to the initial situation when international involvement was needed to protect the Kosovo Albanians from Serbian oppression, now the foreign presence was expected to protect the Serbs from the Kosovo Albanians. This was even more pressing given the Albanian position from the very beginning: they accepted nothing less than full independence. Aware of the complexity surrounding the future of Kosovo, the international community did not want to leave an impression that they would tolerate further acts of discrimination and violence. Consequently, the policy of 'standards before the status' was inaugurated by the third UN Mission in Kosovo under its chief, Michael Steiner of Germany – a well-conceptualized policy that was later abandoned, when Albanians were deemed unlikely to be able to fulfil the standards in the foreseeable future.

When considering the procedure to determine the future status of Kosovo (in accordance with UNSC Resolution 1244), European officials tended to insist that 'any solution must be fully compatible with European values and norms, comply with international legal instruments and obligations and the UN Charter, and contribute to realising the European prospects of Kosovo and the region' (EU Council 2005: 34). Aware of the complexity of the whole process, the EU asked the parties 'to show goodwill, so as to achieve a mutually acceptable solution' and especially 'the authorities in Belgrade actively to encourage the Serbs of Kosovo to take their place in Kosovar institutions, to exercise their democratic rights there' (EU Council 2005). It is worth noting that by 2006 Javier Solana and Olli Rehn had already expressed their support for Kosovo's independence by recommending three courses of action that would assist Kosovo to become a reliable partner with an EU perspective: first, to create a post holding a twofold mandate (to lead the international community's work in the region and to serve as the EU special representative to Kosovo), second, to launch a new EU mission, under the European Security and Defence Policy (to help reform and strengthen Kosovo's justice system), and, finally, to use financial instruments to help Kosovo prepare for the Stabilisation and Association Agreement (Tigner 2006).

In February 2008, the General Affairs and External Relations Council agreed to establish the EU Rule of Law Mission in Kosovo (EULEX), including political and judicial personnel, both international and local, with the task of monitoring and advising. However, there was not much that the EULEX could have immediately done. As noted in the later official report, Kosovo was trying to make some progress, but without any significant success. For example, 'the judicial system remain[ed] weak at all levels' and '[c]orruption was still widespread' (European Commission 2008d: 15), there was 'a lack of capacity to implement and upgrade human rights standards' (European Commission 2008d: 17), money laundering and drug trafficking continued to be 'a very serious problem' and so on (European Commission 2008d: 53). The later reports noted some additional progress, but still not enough to give the impression that Kosovo would be able to secure EU candidate status soon. However, the European Commission reconfirmed the relevance of the previously adopted resolution on Kosovo by the European

Parliament, encouraging 'EU Member States to step up their common approach towards Kosovo' (meaning that all of them should recognize its independence) and underlining that 'the prospect of accession to the EU is a powerful incentive for the necessary reforms in Kosovo' (European Commission 2010a: 5).

For the Serbs (in Serbia, Bosnia and Herzegovina and Kosovo) it has been difficult to accept that Kosovo might be lost forever. Such a feeling was further intensified when the International Court of Justice (ICJ) delivered its advisory opinion in July 2010, concluding that Kosovo's declaration of independence of the 17 February 2008 was in accordance with international law and did not violate UN Security Council Resolution 1244 (ICJ 2010). Still, the Serbian authorities have not given up on their southern province, hoping that some new disclosures, such as Dick Marty's December 2010 report about the involvement of members of the Kosovo Liberation Army in human organ trafficking, currently under investigation by EULEX (Council of Europe 2010), might change the current state of affairs in Serbian favour.

Additional Challenges and Questions

Since the collapse of the Yugoslav federation in 1992, the post-Yugoslav space has received an immense amount of attention from Europe, both its state and non-state actors. Often, lengthy analyses have focused on the lessons learnt and what kind of policies would be suitable for the Balkan region, leaving an impression that there is no alternative to EU integration. In contrast to the Slovenes, who openly insisted on their pro-European orientation in order to secure international recognition and then fulfilled the Copenhagen criteria in order to join the EU in 2004, other Yugoslav republics have seemed to be less aware of and committed to their European perspective. As observed in the literature:

> [t]he only "alternative" to EU integration and democratic consolidation remains continued democratic stagnation and delays to further integration. This particular dynamic appears often in the interest both of some EU member states reluctant to support further EU enlargement and some elites in the Western Balkans, who feel compelled to advocate EU accession formally due to high levels of popular support for European integration, but who have little incentive to pursue this agenda vigorously for fear of undermining their own power base. (Bieber 2012: 8)

As the above overview of the progress of individual post-Yugoslav states has demonstrated, the speed of democratization and Europeanization has been asymmetrical, and something that works in one of the states will not necessarily work in others. This, of course, creates additional concerns within the Brussels administration as to which mechanisms are likely to produce the most appropriate and timely results. Thus, it is not only a matter of the number of chapters and reforms to be addressed and implemented, it is also about choosing the most suitable strategies to convince the locals of the benefits of EU membership rather than opting for policies that would alienate them further. It is also understandable that some political parties and the general public have doubts about the EU's intentions; after all, Europe was initially involved in helping to dismember the SFRY, and, since the 1999 Kosovo crisis, has offered support to the post-Yugoslav space, but on the EU's own terms.

The following chapters present a variety of issues characterizing the links between the EU and individual post-Yugoslav states. Looking at two decades of dynamics, it is clear that such links have often encountered different challenges and obstacles, generated by both the Europeans and the locals. While every obstacle is accompanied by a delay in EU integration, it also tells us more about the very nature and functioning of the Union and its potential members.

References

Almond, M. 1994. *Europe's Backyard War: The War in the Balkans*. London: Heinemann.

Bieber, F. 2012. The Western Balkans Are Dear – Long Live the Balkans! Democratisation and the Limits of the EU, in *Unfinished Business: The Western Balkans and the International Community*, edited by V. Džihić and D. Hamilton. Washington, DC: Center for Transatlantic Relations, 3–9.

Boduszyński, M.P. 2010. *Regime Change in the Yugoslav Successor States: Divergent Paths towards a New Europe*. Baltimore, MD: Johns Hopkins University Press.

Braniff, M. 2011. *Integrating the Balkans: Conflict Resolution and the Impact of EU Expansion*. London and New York, NY: I.B. Tauris.

Chomsky, N. 1999. *The New Military Humanism: Lessons from Kosovo*. London: Pluto Press.

Council of Europe 2010. Inhuman Treatment of People and Illicit Trafficking in Human Organs in Kosovo. 12 December 2010. AS/Jur(2010)46.

Ćurak, N. and Turčalo, S. 2012. Bosnia and Herzegovina: The International Administration of a Captured Country, in *Unfinished Business: The Western Balkans and the International Community*, edited by V. Džihić and D. Hamilton. Washington, DC: Center for Transatlantic Relations, 61–79.

Daalder, I.H. and O'Hanlon, M.E. 2000. *Winning Ugly: NATO's War To Save Kosovo*. Washington, DC: The Brookings Institution.

Danforth, L.M. 1995. *The Macedonian Conflict: Ethnic Nationalism in a Transnational World.* Princeton, NJ: Princeton University Press.

Dragović-Soso, J. 2007. Why Did Yugoslavia Disintegrate? An Overview of Contending Explanations, in *State Collapse in South-Eastern Europe: New Perspectives on Yugoslavia's Disintegration*, edited by L.J. Cohen and J. Dragović-Soso. West Lafayette, IN: Purdue University Press, 1–39.

Džihić, V. 2007. Prospects for the Europeanisation of State-Building Efforts in Kosovo and Bosnia. *A Quarterly Newsletter on German and European Foreign Policy*, 8(23), 24–31.

EPC 1991a. Declaration on Yugoslavia. *Press release*. Brussels, 27 August 1991.

EPC 1991b. Declaration on Yugoslavia. *Press release*. Brussels, 18 October 1991.

EPC 1992a. Statement on Bosnia and Herzegovina. *Press release*. Brussels, 11 April 1992.

EPC 1992b. Statement on Bosnia and Herzegovina. *Press release*. Brussels, 16 April 1992.

EPC 1992c. Declaration on the Former Yugoslav Republic of Macedonia. *Press release*. Brussels, 4 May 1992.

EPC 1992d. Declaration on the Situation in Yugoslavia. *Press release*. Brussels, 15 June 1992.

EPC 1992e. European Council Declaration on Former Yugoslavia. *Press release*. Brussels, 16 October 1992.

EU Council 1993. *Presidency Conclusions*. Copenhagen, 21–22 June 1993. SN 180/1/93 REV 1.

EU Council 2003. *Thessaloniki European Council, 19–20 June 2003: Presidency Conclusions*. Brussels, 1 October 2003. 11638/03.

EU Council 2004. *Presidency Conclusions*. Brussels, 19 July 2004. 10679/2/04 REV 2.

EU Council 2005. *Presidency Conclusions*. Brussels, 15 July 2005. 10255/1/05 REV 1.

EU Council 2008a. *Presidency Conclusions*. Brussels, 14 February 2008. 16616/1/07 REV 1.

EU Council 2008b. *EU Signs Stabilisation and Association Agreement with Bosnia and Herzegovina*. Brussels, 16 June 2008. 10748/08 (Presse 180).

EU Council 2008c. *Report on the Slovenian Presidency of the EU Council*. Ljubljana, 3 July 2008. Available at: http://www.svez.gov.si/fileadmin/svez.gov.si/pageuploads/docs/predsedovanje_eu/REPORT_ON_THE_SLOVENIAN_PRESIDENCY_OF_THE_EU_COUNCIL.pdf [accessed: 15 January 2012].

European Commission 1992. *XXVI General Report on the Activities of the European Communities*. Luxemburg: Office for Official Publications of the European Communities.

European Commission 1993. *XXVII General Report on the Activities of the European Communities*. Luxemburg: Office for Official Publications of the European Communities.

European Commission 1997a. *Agenda 2000: Commission Opinion on Slovenia's Application for Membership of the European Union*. Brussels, 15 July 1997. DOC/97/19.

European Commission 1997b. *Regional Approach to the countries of South-Eastern Europe*. Brussels, 3 October 1997. Available at: http://ec.europa.eu/

enlargement/pdf/key_documents/9710_report_a_en.pdf [accessed: 1 February 2012].

European Commission 1999. *Regular Report from the Commission on Slovenia's Progress towards Accession*. Brussels, 13 October 1999. Available at: http://ec.europa.eu/enlargement/archives/pdf/key_documents/1999/slovenia_en.pdf [accessed: 15 January 2012].

European Commission 2002a. *Former Yugoslav Republic of Macedonia: Stabilisation and Association Report*. Brussels, 4 April 2002. SEC(2002)342.

European Commission 2002b. *Regular Report on Slovenia's Progress towards Accession*. Brussels, 9 October 2002. SEC(2002)1411.

European Commission 2003. *Bosnia and Herzegovina: Stabilisation and Association Report 2003*. Brussels, 26 March 2003. SEC(2003)340.

European Commission 2004a. *Former Yugoslav Republic of Macedonia: Stabilisation and Association Report*. Brussels, 12 March 2004. SEC(2004)373.

European Commission 2004b. *Bosnia and Herzegovina: Stabilisation and Association Report 2004*. Brussels, 30 March 2004. SEC(2004)375.

European Commission 2004c. *Communication from the Commission. Opinion on Croatia's Application for Membership of the European Union*. Brussels, 20 April 2004. COM(2004)257 final.

European Commission 2006. *The Former Yugoslav Republic of Macedonia 2006 Progress Report*. Brussels, 8 November 2006. SEC(2006)1387.

European Commission 2007. *Bosnia and Herzegovina 2007 Progress Report*. Brussels, 6 November 2007. SEC(2007)1430.

European Commission 2008a. *Communication from the Commission to the European Parliament and the Council. Western Balkans: Enhancing the European Perspective*. Brussels, 5 March 2008. COM(2008)127 final.

European Commission 2008b. *The Former Yugoslav Republic of Macedonia 2008 Progress Report*. Brussels, 5 November 2008. SEC(2008)2695.

European Commission 2008c. *Montenegro 2008 Progress Report*. Brussels, 5 November 2008. SEC(2008)2696.

European Commission 2008d. *Kosovo (under UNSCR 1244/99) 2008 Progress Report*. Brussels, 5 November 2008. SEC(2008)2697.

European Commission 2009a. *Croatia 2009 Progress Report*. Brussels, 14 October 2009. SEC(2009)1333/2.

European Commission 2009b. *Montenegro 2009 Progress Report*. Brussels, 14 October 2009. SEC(2009)1336.

European Commission 2009c. *Bosnia and Herzegovina 2009 Progress Report*. Brussels, 14 October 2009. SEC(2009)1338.

European Commission 2009d. *Serbia 2009 Progress Report*. Brussels, 14 October 2009. SEC(2009)1339.

European Commission 2010a. *Kosovo* 2010 Progress Report*. Brussels, 9 November 2010. SEC(2010)1329.

European Commission 2010b. *Serbia 2010 Progress Report*. Brussels, 9 November 2010. SEC(2010)1330.

European Commission 2010c. *The Former Yugoslav Republic of Macedonia 2010 Progress Report*. Brussels, 9 November 2010. SEC(2010)1332.

European Commission 2010d. *Communication form the Commission to the European Parliament and the Council. Commission Opinion on Montenegro's Application for Membership of the European Union*. Brussels, 9 November 2010. COM(2010)670.

European Commission 2011a. *Montenegro 2011 Progress Report*. Brussels, 12 October 2011. SEC(2011)1204.

European Commission 2011b. *Bosnia and Herzegovina 2011 Progress Report*. Brussels, 12 October 2011. SEC(2011)1206.

European Commission 2011c. *Communication from the Commission to the European Parliament and the Council: Enlargement Strategy and Main Challenges 2011–2012*. Brussels, 12 October 2011. COM(2011)666 final.

European Parliament 1977. Debate No. C 223/36, *Journal officiel des Communautés européennes*. Brussels, 14 November 1977.

Farkas, R.P. 2007. *Democratization in the Balkans: Prescription for a Badly Scarred Body Politic*. Boston, MA: Northeastern University Press.

Government of the Republic of Croatia 1999. *Action Plan for European Integration*. Zagreb: Office for European integration.

Government of the Republic of Serbia 2004. *Evropska orijentacija gradjana Srbije*. Belgrade: Serbian Government Office for EU Integration.

Government of the Republic of Serbia 2012. *Serbia's EU Integration Supported by 51% of Citizens*. Belgrade: Serbian Government Office for EU Integration.

Gow, J. 1997. After the Flood: Literature on the Context, Causes, and Course of the Yugoslav War-Reflections and Refractions. *Slavonic and East European Review*, 75(3), 446–84.

Gow, J. and Carmichael, C. 2010. *Slovenia and the Slovenes: A Small State in the New Europe*. London: Hurst and Company.

Gowan, P. 2000. The Euro-Atlantic Origins of NATO's Attack on Yugoslavia, in *Masters of the Universe? NATO's Balkan Crusade*, edited by T. Ali. London: Verso, 3–45.

Hammel, E.A. 1999. The Bombing of Serbia: Back to the Future. Available at: http://www.demog.berkeley.edu/~gene/bombserbs2.html [accessed: 20 January 2012].

ICJ 2010. *Accordance with International Law of the Unilateral Declaration of Independence in Respect of Kosovo*. 22 July 2010. General List No.141.

Investment Development Consultancy/Development Strategies. 2001. *Evaluation of EC Country Strategy: FYR Macedonia 1996–2001*. Available at: http://ec.europa.eu/europeaid/how/evaluation/evaluation_reports/reports/cards/951608_en.pdf [accessed: 5 February 2012].

Job, C. 2002. *Yugoslavia's Ruin: The Bloody Lessons of Nationalism*. Lanham, MD: Rowman and Littlefield.

Jović, D. 2001. The Disintegration of Yugoslavia: A Critical Review of Explanatory Approaches. *European Journal of Social Theory*, 4(1), 101–20.

Jović, D. 2011. Turning Nationalists into EU Supporters: The Case of Croatia, in *The Western Balkans and the EU: 'The Hour of Europe'*, edited by J. Rupnik. Paris: EU Institute for Security Studies, 33–45.

Kent, S.A. 1997. Writing the Yugoslav Wars: English Language Books on Bosnia (1992–1996) and the Challenges of Analyzing Contemporary History. *American Historical Review*, 102(4), 1085–114.

Knaus, G. and Cox, M. 2004. Bosnia and Herzegovina: Europeanisations by Decree?, in *The Western Balkans: Moving On*, edited by J. Batt. Paris: EU Institute for Security Studies, 55–68.

Kučan, M. 1996. Slovenia Signs an Association Agreement with the European Union. 10 June 1996. Available at: http://www2.gov.si/up-rs/1992-2002/mk-ang.nsf/0/e8e6ce21b83d5f9bc12567900059e03a?OpenDocument [accessed: 16 January 2012].

Leatherman, J., DeMars, W., Gaffney P.D. and Väyrynen, R. 1999. *Breaking Cycles of Violence: Conflict Prevention in Interstate Crises*. West Hartford, CT: Kumarian Press.

Longinović, T. 2011. *Vampire Nation: Violence as Cultural Imaginary*. Durham, NC: Duke University Press.

Marović, S. 1999. Miloševićeva Srbija uplašeno ćuti. *Vreme*, No. 455, 25 September 1999.

Morrison, K. 2009. *Montenegro: A Modern History*. London and New York, NY: I.B. Tauris.

Nakarada, R. 2008. *Raspad Jugoslavije: Problemi tumačenja, suočavanja i tranzicije*. Beograd: Službeni glasnik.

Noutcheva, G. 2007. *Fake, Partial and Imposed Compliance: The Limits of the EU's Normative Power in the Western Balkans*. Brussels: Centre for European Policy Studies.

Ordanoski, S. 2011. The Story of Macedonian Populism: 'All we want is everything!', in *The Western Balkans and the EU: 'The Hour of Europe'*, edited by J. Rupnik. Paris: EU Institute for Security Studies, 95–109.

Radeljić, B. 2012. *Europe and the Collapse of Yugoslavia: Role of Non-State Actors and European Diplomacy*. London and New York, NY: I.B. Tauris.

Radulović, M. 2011. Montenegro's Journey Towards EU Accession, in *The Western Balkans and the EU: 'The Hour of Europe'*, edited by J. Rupnik. Paris: EU Institute for Security Studies, 79–93.

Ramet, S.P. 2005. *Thinking about Yugoslavia: Scholarly Debates about the Yugoslav Breakup and the Wars in Bosnia and Kosovo*. Cambridge: Cambridge University Press.

Ramet, S.P. and Pavlaković, V. 2005. *Serbia since 1989: Politics and Society under Milošević and after*. Seattle, WA: University of Washington Press.

Ristić, I. 2009. Serbia's EU Integration Process: The Momentum of 2008, *Panoeconomicus*, 1, 111–25.

Rupnik, J. 2004. The Demise of Balkan Nationalisms? A Skeptical View, in *The Western Balkans: Moving On*, edited by J. Batt. Paris: EU Institute for Security Studies, 99–110.

Shah, A. 1999. Did NATO really Act out of Humanitarian Concerns? *Global Issues*, 27 March.

Smith, C.J. 1996. Conflict in the Balkans and the Possibility of a European Union Common Foreign and Security Policy, *International Relations*, 13(2), 1–21.

Stokes, G., Lampe, J., Rusinow, D. and Mostov, J. 1996. Instant History: Understanding the Wars of Yugoslav Succession. *Slavic Review*, 55(1), 136–50.

Tigner, B. EU Ministers Support Kosovo Independence. *Defense News*, 18 July 2006. Available at: http://www.defensenews.com/story.php?F=1961498&C=europe [accessed: 5 November 2011].

Tocci, N. 2007. *The EU and Conflict Resolution: Promoting Peace in the Backyard.* Oxon: Routledge.

Tzifakis, N. 2012. Bosnia's Slow Europeanisations, *Perspectives on European Politics and Society*, 13(2), 131–48.

UN 1996. *UN Security Council Resolution 1058.*

UN 1999. *UN Security Council Resolution 1244.*

UNMIK 2003. Djindjić wants status resolved now, Koštunica in five years. Available at: http://www.unmikonline.org/press/2003/mon/jan/lmm200103.htm#4 [accessed: 7 November 2011].

Vujačić, I. 2011. Deset godina političke tranzicije u Srbiji. *Analiza politike*, Summer 2011, 18–23.

Chapter 2

Returning to Europe: Post-Yugoslav Europeanization of Slovenia and the Balkan *Other*

Damjan Mandelc

This chapter goes beyond argumentation and explanation of the key circumstances that led to the breakup of the Socialist Federal Republic of Yugoslavia (SFRY). However, there are plenty of reasons and a number of (in)direct effects that have contributed to the development of the Slovene nation-state in the last two decades. By employing some convincing models we can reliably describe the Slovene nation-formation practices of (ethno)nationalism and ethnocentrism (focusing on Roma, Muslim, former SFRY nationals and minorities, the Erased and so on) walking hand in hand with post-national and transnational tendencies, as well as the Europeanization processes. While gaining full EU membership, joining the common currency and becoming an active member of the EU, Slovenia temporarily and intentionally distanced itself from the common former space (and identities), and has only recently started to reinforce its presence in the space of the ex-Yugoslav state(s). In addition to analysis and reflection on the past 20 years, the author's research question is whether there is any prospect of re-establishing stronger cooperation among the former Yugoslav states, and also what the potential is for future strategic partnership of the countries of the former Yugoslavia in the light of the current and future constellations of power and development in the EU.

What If?

Let's imagine a parallel reality: The United States of Yugoslavia (USY) has just celebrated 20 years since its emergence in 1991. After unsuccessful attempts to gain independence, Slovenia and Croatia proposed to the rest of the constitutive units of the decomposing socialist federal state that they should peacefully transform into a strong political union, based on the combination of the existing model of multinational and multi-ethnic states from around the world. Their ambition and strongest argument was that of avoiding the high risk of war. While the newly awakened Serbian ethnic nationalism, led by Slobodan Milošević, seriously threatened to ignite the war flame, the other republics, together with the emerging civil society and the political opposition in Serbia, decided to search for solutions and compromises. Fortunately, reason prevailed. Slovenia, Croatia, Bosnia and Herzegovina, Montenegro, Vojvodina, Serbia, Kosovo and Macedonia formed a progressive political alliance, and the newly formed institutions moved to different state capitals: Ljubljana got the seat of the Federal Parliament, Zagreb the Constitutional Court, the President of the United States of Yugoslavia's headquarters remained in Belgrade while the army headquarters moved to Priština, and the Upper House of Parliament to Skopje. The USY adopted a strong democratic constitution with parliamentary democracy and the rule of law. Previously oppressed oppositions and representatives of the former regime achieved reconciliation, all political prisoners were released and the armed forces underwent a reduction in their numbers. Economically, the country maintained a strong social model based on a system of self-management and strong unions; a solid balance between economic development, based on export, knowledge and research, and social egalitarianism was created. The USY moved from being a major player in the Non-Aligned Movement to now playing a lead role in the newly formed League of the Aligned, an international political organization that promotes and implements a *new global deal*, a sustainable, democratic, egalitarian model of society and international relations, with a strong green and social agenda. Its member states include most of Latin and Central America, most of Africa, most of Asia, together with some post-Soviet republics, and some non-EU countries.

Let's imagine a further twist: In 2010, a group of two hundred historians from all over the world carried out a comprehensive projection study of what might have happened if the Socialist Federal Republic of Yugoslavia had really fallen apart and if agreement among the republics had not been reached. The title of the 2011 best-seller (to be found in the political fiction section of bookshops) is *Counting the Killed*, estimating 500 victims in Slovenia, and then 20,000 in Croatia, 200,000 in Bosnia and Herzegovina and 15,000 in Kosovo, if war had erupted in Yugoslavia. According to the historians, the outbreak of war would have been inevitable, since it is reasonable to expect that the Yugoslav army would have cruelly suppressed any attempt of secession, following the lead of xenophobic and intolerant nationalist leader Slobodan Milošević. To prevent communications, the Yugoslav army would have attacked Slovenia first. Then, if Croatia had declared independence, Serbia would have wanted to prevent the loss of almost half a million Serbs living in Croatia. A devastating scenario would have developed in even more ethnically mixed Bosnia and Herzegovina; it would have been a situation in which any attempt to split the state of Bosnia along ethnic borders would have led to a long and bloody war with no winner and hundreds of thousands of victims. And if the speculation as to what might have happened in Kosovo is correct, the Serbs and the Yugoslav army would have tried to prevent the loss of Kosovo at any cost. It might even have reached a level where international military forces intervened to prevent the killing of civilians. Yugoslavia would have drowned in blood, the historians claim. Luckily, Milošević was replaced by a democratic political figure and returned to work as a banker. The power of the army was reduced and negotiations over the future of the federal state began at once.

Back to the Facts

It was months before the independence referendums that the republics of Slovenia and Croatia made an offer (first on 3 October 1990 and then again on 16 October 1990) to the Yugoslav Presidency to create a Yugoslav confederation. Both times the proposals were rejected and one can only speculate what would have happened if the federal units, federal government and military had initiated a dialogue instead of searching for reasons for war and for the breakup. Slovene political leadership, based on the broad democratic support of all the relevant segments of society (political and economic elites, civil society and population), decided to continue with activities for the independence referendum. On 23 December 1990, Slovenes voted to decide whether Slovenia should become an independent state. The vast majority was in favour of independence.

At the close of 2011, Slovenia celebrated its twenty-first anniversary of statehood. Two decades are not enough to provide a valid assessment of epochal breaks and transformations since the early 1990s. However, there are now new generations of children born in independent Slovenia, who never experienced Yugoslavia and for whom Yugoslavia is only a history-book reference. Their narrative is different from their parents' and grandparents' narrative, their symbols and political identity are completely different from the identities of the pre-1990s generations. The author of this chapter (born in 1977) happened to experience them all: two political and economic systems, different currencies (Yugoslav dinar, Slovene tolar, and now euro), diverse symbols, and so on. It is impossible to describe the intense and turbulent political and social period and the multiple transformations of a nation in one short chapter. It is, however, possible to delineate the frames within which society moved, the conditions and consequences of its development, to evaluate the key phases and breaks, to take perspective on relevant social problems, and to anticipate possible future developments on the basis of the current situation.

Yugoslavia, being a multinational state, witnessed tensions between two-level loyalties: loyalty to a federal state on the one hand, and loyalty to one's own national community on the other. Keeping the balance between the two was one of the most challenging tasks in Yugoslavia, a task that central authorities could not handle. As we read in Matić (2006: 267), communism determined the identity of Yugoslavia, according to which, loyalty to the state equalled loyalty to communist ideology.[1] However, with each passing year, the communist ideal started to lose its appeal. National (communist) elites could not find a common

1 Walker Connor writes that in order to maintain a multinational state, one must keep the balance between the two types of loyalty, loyalty to the nation and loyalty to the state

language for how to resolve the crisis; during the turbulence of the late eighties and early nineties, a shift from Yugoslav (transnational, unitary, socialist ideology and identity) towards (ethno)nationalistic sentiments took place.

If we identify the crucial circumstances that accompanied the transition from Yugoslavia to independent Slovenia, we can note the rather stable economic situation of socialist Slovenia – while Slovenes made up only 8 per cent of the total population, their economy contributed one-third to Yugoslav export and comprised a fifth of the GDP of the whole country. The Slovene GDP per capita in 1975 was 201 per cent of Yugoslav GDP per capita (Tomc 1993: 222). In 1985, the GDP per capita in Yugoslavia and Serbia was at the level of Turkey, Kosovo at the level of Pakistan, Croatia at the level of Greece and Slovenia at the level of Spain and New Zealand (Borak 2000: 322–3). All in all, Yugoslavia was once a regional industrial power and economic success; from 1960 to 1980 the average GDP growth was 6 per cent, medical care and education were free, life expectancy was 72 years (World Bank 1991). Economically and politically, as its most developed region, Slovenia had a rather significant impact on Yugoslavia. For all these reasons, Slovenia did not look for its place outside Yugoslavia until the late eighties (Repe 2001: 6) when the economic situation worsened and differences in development had further grown.

When the external geostrategic circumstances changed and communism all over Central and Eastern Europe faced collapse, Yugoslavia reached a rather difficult situation. Its inner contradictions, economic inefficiency and ethnic (also religious) tensions began to surface. These circumstances have severely affected the new-born Slovene state and forced it to undertake major structural changes in order to move towards the developed economies of Western Europe. As a small economic space, Slovenia started its transition rather well prepared, because the process of transformation from the centrally planned socialist economy to a market economy had started some years before the final collapse of the socialist system. However, the transition to a market economy did not occur without any consequences; like all former socialist states, Slovenia also fell into recession, with severe consequences (negative economic growth, high unemployment, inflation, substantial economic losses, and general illiquidity). Among the biggest problems independent Slovenia had to face was the loss of the internal market. In the two years after independence Slovenia lost approximately 76 per cent of the former Yugoslav market. Moreover, it also lost some of the markets in the Eastern European countries, and the whole situation was accompanied by the great recession in the Western Europe in the early nineties. The consequences of the situation for Slovenia were high unemployment, price rises, and a lower standard of living.

– thus, in a situation where these two loyalties come into conflict, loyalty to the state loses the battle (Connor 1994: 81).

Processes of Democratization of Slovenia

The most sensitive of times, when democracy is facing the challenge of lack, is the transition between different political and social regimes, a time of rapid socioeconomic change, and political, religious and cultural ferment (Alvis 2005). Slovene democratization, in its rather short history as an independent state, was widely perceived as non-problematic; in the broader transnational political arena, the country's move towards independence was even used as a model, since it did not result in the wars that befell other states of the collapsing Yugoslav region (Bebler 2010, Fink-Hafner 1995, Glenny 1996, Janjić 1992, Ramet 1996).

The processes of democratization in Slovenia were largely accompanied by the process of re-nationalization: Slovenia followed the 'ethnic' model of state building. Despite emerging democratic structures, (certain) minority groups in the newly formed state were being exposed to discrimination. One can find proof for such discrimination in several cases; and while Italian and Hungarian minorities were widely financed and supported by the state, some other groups were not as generously awarded.[2] In giving a proper civic status to the autochthonous 'old' minorities, Slovenia *forgot* about other groups, specifically those belonging to other Yugoslav republics. The implied assumption was that by granting protection to the existing legally recognized minorities, the system would block any possible ethnic conflicts in the future, while human rights would be observed.

Therefore, when speaking of minority issues, political and legal discourse focused only on the pre-existing minorities, called the 'old', indigenous, or 'autochthonous'.[3] As Rizman reminds us, 'the Council of Europe's Commissioner for Human Rights, Alvaro Gil-Robles during his visit in 2003 already suggested to the Slovene government to get rid of distinction between autochthonous and non-autochthonous national minorities … it can still be said that all Slovene governments ignored the issue and just mentioned advice' (Rizman 2006: 129).

Article 5 of the Slovene Constitution specifically names only the Italian and Hungarian minorities as protected national minorities. Only once is the word

---

2 For instance, an unofficial Muslim (predominately of Bosnian origin) minority has been now waiting for almost four decades to be granted permission by state authorities to build the first mosque in Slovenia. As the same time, the Serbian (again unofficial) ethnic/religious minority waited for more than two decades to obtain permission to build the orthodox cultural centre in the capital of Slovenia. Not to mention the obstacles for Roma communities, whose minority status was excluded from the Slovene Constitution (which provided only the status of ethnic community), and while the Constitution required the 'umbrella law' on Roma ethnic community, such law was only adopted in March 2007. On the situation of the Roma community, see Republic of Slovenia, Office for National Minorities, http://www.uvn.gov.si/en/minorities/roma_community/.

3 Due to its arbitrary nature, the term 'autochthonous' is no longer used in EU documents with regard to national and ethnic minorities. In this sense, Slovenia is an exception (Kralj 2009: 2).

'minorities' (*manjšine*) actually used in the Constitution, and it is in specific reference to the autochthonous minorities. Not once does the Constitution mention the protection of any other minorities, such as, for example, minorities from the former republics of Yugoslavia. Article 64 of the Constitution protects the right of autochthonous minorities to use national symbols and keep national identity cards; it grants them the right to education and schooling in their own languages. These rights are not guaranteed to the new minorities; legally invisible, the new minorities, therefore, neither exist in the political vocabulary, nor do they appear in state institutions. However, as we will see below, they do have a place in the national imaginary, where they play an ambiguous role of the (un)wanted residue from the Yugoslav history.

Discrimination as a form of ethnocentrism has been prevalent throughout the history of independent Slovenia. The presence of inhabitants and workers from other former Yugoslav republics, who were perceived as remainders of the common Yugoslav history, was seen as an instrument of 'Yugoslavization' of Slovenia, from which the country wanted to distance itself as much as possible. Slovenia has sought distance from its Balkan neighbourhood, and '[has returned] home, to the company of western European nations, to where it has always belonged, by virtue of culture, history and civilization' (Petrović 2009: 10). All traces of a former shared Yugoslav history thus needed to be removed from society for it to successfully make the transition into a modern European national state. Europeanization was thus seen as separation from the Yugoslav legacy, and not as a possible result of democratization.

In his article on nations, identity, and conflict, Jonathan Glover writes that 'nationality is often thought of as something "natural" or presocial ... naturalness is reinforced by stories nations often have about their own antiquity. But some historians and social scientists emphasize the relative modernity of the European nation-states, dating them from around the end of the eighteenth century' (Glover 1997: 12). We see a replica of this nineteenth-century model reiterated in the Slovenia of the late twentieth century. Driven by the collective enthusiasm for forming an independent state, national elites in Slovenia exaggerated the need for homogeneity. The discourse they employed stressed the need to disassociate the society culturally and historically from its recent, post-Second-World-War socialist past and embark on a path of development which would eradicate traces of the multicultural and multi-ethnic formula of collective identity building. The immigrants from the former Yugoslavia, being a historical reminder of the past that once Slovenia was consensually part of, began to play the convenient role of the ethnic 'fifth column' (Kymlicka 2007) within the emerging independent national society. By their presence, immigrants from former Yugoslavia of multi-ethnic, transcultural and transborder 'brotherhood and unity' destabilized the certainty of the nationalist discourse, which claimed that ethnocultural homogeneity could be the only proper base for national loyalty. To grant such people legal protection would not only mean granting them public visibility but, at the symbolic level, it was also an acceptance of Yugoslav history as a moment in the identity building of the hegemonic Slovene national citizen. Therefore, obliteration of the fact of the very existence of immigrants and the need for their legal protection as minorities conveniently served several purposes at once. The symbolic neglect of their legal group status may not have been the most violent one in terms of their political representation, but it most certainly was the most forceful in terms of public legitimation of their exclusion from the bounds of the national society.

We can agree with Smith (1981: 29) and argue that aggressive nationalism is no less a potent tool for mobilizing social fears in economically advanced states than is the case in underdeveloped countries. In fact, Slovene political elites skilfully employed popular perceptions of the continuing economic exploitation of vital economic powers of the North by the South in post-Tito Yugoslavia. Although this is not the place to discuss how the uneven development in Yugoslavia has actually helped the national political and economic elites to maintain their positions of power, it is important to link this ideological stigmatization of the immigrants as the prolongation of this past practice of exploitation *within* the now independent, post-Yugoslav state.

To sum up, the international community perceived the process of democratization and state building in Slovenia as an exemplary model of transition into the new social and political order. But while the commitment to form a democratic state based on the protection of human rights and the enacted legislation was publicly and legally articulated, in many respects the reality was far from the declared model. Daniele Conversi writes that '[a]ny process of nation-building insensitive to ethnic nuances and local subjectivities implies a parallel process of *nation-destroying* among minority groups' (2003: 3). This statement can be used to describe the process in Slovenia, namely, the process of state building founded in parallel construction and destruction of legacies, which made the immigrant minorities a legitimate and autonomous (civic) national subject. While Slovene political elites turned to pre-First-World-War history to revive the allegedly pristine and authentic ethnic substance of the nation, they blatantly omitted any cultural traces of collective identity that could be considered as Yugoslav. It is only by considering this wider political and ideological context of transition towards democracy that the collapse of the basic democratic mechanisms of protection of the legal state, and the violation of human rights can be explained (Mandelc and Učakar 2011: 28–32).

The emergence of (ethnic) nationalism made some large holes in the Slovene democracy. To explain some of these problems and place them in their socio-historical context, we cannot avoid a brief theoretical reflection on the key concepts that characterize state building; namely human rights, nationalism, and the theory and practice of citizenship.

Human Rights

In principle, human rights belong to all individuals with their birth, but they can only be granted by a sovereign nation-state. The result is that citizenship is the only basis for accessing human rights in a political community. Prevention of statelessness is one of the main principles in the international law that states are obliged to follow when determining their civic body, since it is difficult to ensure human rights to those who find themselves without any kind of citizenship or permanent resident status. Nation-states are thus the sole political entities that can provide space where human rights can be exercised (Arendt 2004). Defining the civic body of the state on one hand has always meant mass exclusion on the other. This way, modern nations have established territorially bounded national sovereignty, achieving internal homogeneity through dialectical definition of an otherness that was excluded, dehumanized and left without any rights. Establishing imperial sovereignty thus implies mutual normalization and integration of exclusionary practices (Kurnik 2007).

The newly formed Slovene civic body, which was at the time of the creation of the independent state defined by ethnic measures, left out those individuals that were not perceived as ethnically Slovene. This had a great impact on many groups within Slovenia, especially the group of the Erased. Since such exclusive principles and practices are often connected to nationalistic doctrine we should take a closer look into the theories of nationalism.

(Ethno)nationalism

Nationalism has a relatively ambivalent nature; it may be characterized as a political movement willing to achieve protection of collective cultural and political rights, autonomy or self-determination, state ideology, or (most often) as an extreme emotion of hatred against foreigners, cultural exclusivism and xenophobia. The reasons for such plurality of perception lie in the inability of scholars to distinguish between the two dimensions of nationalism. On the one hand we have 'traditional' nationalism that argues for a state/country for every nation; the nation-state has to belong to its ethnocultural or political group to protect its interests (culture, traditions, heritage and territory) and enable it to flourish. This form of nationalism is perceived as 'classic nationalism', and thrived most strongly in nineteenth-century Europe, later spreading around the world as epidemics and marking some contemporary nationalisms even today. That was the case of the Slovene nationalism in the 1990s, when the state formation was based on the nation-state perception, the newly formed state being perceived as the protector of the majority ethnic group, the Slovenes. Smith (2001) clearly states that a nation is not a state, because the concept of a state refers to an institutional, structural activity, while nation indicates the type of community. Similarly, according to Smith, the nation is not an ethnic community (although they partly overlap, being members of the same family of phenomena, that is, collective cultural identities) because 'the ethnic community usually has no political referent, and in many cases lacks a public culture and even a territorial dimension, since it is not necessary for an ethnic community to be in physical possession of its historic territory' (Smith 2001: 12).

John Breuilly says that 'nationalism can provide identity in a time of rapid change' (Breuilly cited in Özkirimli 2000: 49). It was exactly a period of rapid change that Slovenia was experiencing as it strove to eradicate its Yugoslav past and become a member of the European Community within only a couple years. For decades, Slovenia had longed to return to its place in central Europe. Reconciling with *Mitteleuropa* also meant growing distant from the Balkans. In his first speech to the UN, the first President of Slovenia, Milan Kučan (1992), referred to the Balkans as a region 'with its great backwardness in terms of the progress of civilization and aspirations of the world, its inability to follow the process of emancipation in Eastern Europe'. With this and other similar public statements we can clearly observe that Slovenes felt rather annoyed by the influence of the Balkan past over the new nation-state presence (the most visible group were workers from the other Yugoslav republics who were characterized as people of poor culture with different values. As was mentioned above, in some ways, they were also seen as instruments of the 'Yugoslavization' of Slovenia). This Yugoslavization was exactly what needed to be removed from society if it was going to make a successful the transition into a modern European nation-state. Besides, Slovenes felt they did not belong to the realm of Eastern Europe. They were probably envious of the success achieved by other western European states and the fact that they were excluded from this integration. And finally, an ethnic revival would be an obvious response to these nationalistic threats. In regard to the ethnic revival in Slovenia, Anthony Smith wrote that 'Slovenia enjoys higher living standards than a preponderantly agricultural, and rather backward, Serb core area. And yet discontent and desire for greater autonomy remain powerful political forces. Ethnic groups in economically advantaged region are just as susceptible to violent ethnic protest movements as those which inhabit backward areas, like Ireland in the last century' (Smith 1981: 29).

Among scholars there are strong debates about when in human history nation and nationalism emerged, whether they truly are ancient phenomena, as primordialists claim, or completely new notions, potentially associated with industrialization, or standardization of language and mass literacy, or maybe a political consequence of the American and French revolutions, as modernists believe. Another theoretical position, ethno-symbolism, offered by Anthony Smith, is the internal critic of modernism, claiming that nationalism and nation-state are phenomena of modernity, except that we can indeed identify some pre-existing criteria (traditions, myths and other ethno-symbolic attributes that existed before the emergence of the modern nation-state). This at-first-glance-unsolvable situation may be demonstrated by Gellner's stressing that 'nationalism is not the awakening of nations to self-consciousness: it invents nations where they do not exist' (1964: 169). Smith responds that 'nations and nationalisms are no more invented than other kinds of culture, social organization or ideology' (1991: 71–2). In a similar way, Smith rejects some other conceptions of nationalism, for instance, the claim that nationalism is 'by nature' an extreme force. He says that this is true to some degree, but we also know some rather moderate and democratic nationalisms

(Czech, Catalan, or Quebecois, for instance), demonstrating that generalization is not appropriate or adequate. In many cases it is true, but there are other nationalist movements that address their appeal to collective, cultural and/or minority human rights. Many see nationalism as the key source of destabilization, and again, in several contexts one can observe such a development, but nationalism is not *per se* responsible for the collapse of states; nationalism usually develops on the ruins of existing states (Smith 1995: 152). The destructive power of nationalism is the other side of the coin, the first side of which is uniting – depending on one's reference frame. According to Smith, modern nationalisms have become what religious communities used to be in the past (nationalism as secularized religion), 'communities of history and destiny that confer on mortals a sense of immortality through the judgment of posterity, rather than through divine judgment in an afterlife' (Smith 1995: 158–9).

As mentioned above, most people connect the idea of nationalism to ethnocentrism and racism, violence and hatred against outside groups (Lichtenberg 1997: 158). In their opinion, nationalism is the evil that must be overcome and replaced with cosmopolitanism. On the other hand, the idea on nationalism offers some positive connotations such as attachment to democracy, community, autonomy, aspirations of freedom and preservation of cultures, languages and traditions. Regardless of its ratings, one can solidly claim that nationalism is (still) powerful; it addresses masses and responds to the aspirations of many communities. In this, nationalism is greatly supported by (rational) arguments, usually referring to (a) the right to self-determination (argument of self-determination), (b) the right to prosperity, arguing that every person needs to identify with some broader community, bigger than their family (prosperity argument), (c) the right to redress those collective injustices that were committed in the past (argument of reparation), (d) the view that the world is a better place if it possesses diverse cultures (plurality argument), and (e) the view that the existence of a certain culture requires its preservation and its flourishing, that every culture has its specific value, so it is legitimate to preserve it (argument of inner value) (Lichtenberg 1997: 160–61). Although Lichtenberg is resisting the arguments of nationalism, she is not successful in her mission since she does not offer any alternative to the nationalist(ic) imagining of the world. She notes that not every culture will reach the final stage of self-determination, the independent state, yet it is, however, possible to use different politics of multiculturalism in multicultural communities (promotion of minority languages and other cultural practices) to achieve peaceful coexistence. In this sense, only multiculturalism is the answer, and the politics of recognition that takes from nationalism what is acceptable and directs it to more cosmopolitan ways (Lichtenberg 1997: 172).

For a long time, Slovene nationalism was perceived mainly in the positive perspective, especially when compared to the nationalisms in other former Yugoslav republics that led to devastating wars, which was not the case in Slovenia. But when the nationalist and even xenophobic attitude towards specific minority groups is examined, one can clearly place Slovene nationalism with ethnic nationalism throughout Eastern Europe: it has played the same dual role – on the one hand it was used to homogenize the nation and played an important role in the strengthening of the national identity, but on the other, it was used to divide between 'true members' and others, especially other Yugoslav nationals.

Citizenship

Some languages distinguish *citizenship* from *nationality*, where *citizenship* stands for 'legal relationship, regulating the totality of mutual rights and duties that may exist between an individual and a State' (Horn 1998), and is referred to domestic legal order. According to Dedić (2003: 25–6), *nationality* stands for 'the legal relationship between an individual and a State in international law'; a state thus grants diplomatic protection to its *nationals*. In languages that use only the term *nationality*, there is often a notion of ethnic affiliation, as is common in Central and Eastern Europe. However, as Dedić observes, the European Convention on Nationality perceives nationality as 'legal relationship between an individual and a State, not reflecting ethnic origin of the individual' (2003: 26).

National citizenship as a paradigm for the interpretation of the relationship between the citizen and the individual non-citizen in the political community envisages that citizen rights overlap human rights. In Europe, the possibility of overcoming a national citizenship system is brought forward by the introduction of European citizenship, whose body is difficult to determine, since it depends on how Europe and the European Union are defined. In addition to this, the Maastricht Treaty of 1992 defined the EU citizen as every individual possessing the nationality of a Member State. EU citizenship thus complements national citizenship and does not replace it. Therefore, the legal function of national citizenship remained unchanged. In this way, instead of exceeding the limits of national citizenship, the European Union created a new discrimination (Balibar 2004, Tuccillo Castaldo 2007).

Despite the idea of Europe overcoming the dividing line between sovereign national states, it seems as if the EU as a political entity included both elements: the nation and the state. Thus, by not overcoming national citizenship systems and expanding the 'right of residency' (*ius soli*), which would have enabled real citizenship on the basis of residing in Europe, the European Union created a new discrimination. Immigrants are differentiated into those from other member countries and those from countries outside the EU, the so-called 'third' countries (Balibar 2004). Now the third-country immigrants are discriminated against not only by the national states in which they settle, but also by the European Union. Tuccillo Castaldo (2007) argues that EU enlargement includes several levels of exclusion. The EU's external borders expand mostly with the aim of monitoring migration flows, while inside the EU supervision over inferior(ized) populations keeps getting harsher. Moreover, citizenship legislation is evolving in a way that means that naturalization processes are increasingly determined by economic-income criteria and at the same time, the *ius sanguinis* principle is becoming more and more relevant. Foreigners have become *second-class citizens* (Balibar 2004: 171), who are often stigmatized, with their stay and activity being strictly controlled.

Citizenship is to be understood within a broader social context and not only as a legal category. Balibar (1994: 724) understands citizenship in a narrow sense as 'full exercise of political rights' and in a broader sense as 'cultural initiative or effective presence in the public space (the capacity to be "listened to" there)'. The two tendencies, to equate citizenship to nationality on one side, and to consider citizenship solely as legal fiction on the other, were differently balanced throughout history. But Balibar warns that the dynamic relation between the citizen and the state is overlooked in both cases and therefore talks about 'worksites of citizenship' (2004: 198), thus exposing the dynamic link between the state and the individual. Balibar argues that it is necessary to find a new type of citizenship that will overcome its present bond to nation-state. Citizenship is namely a process formed by those who invoke the right to be citizens. Also, the citizenship we know today was created through historical battles and struggles that will also be needed in the future. As an alternative to today's citizenship, Balibar introduces the term *droit de cité* (the right to citizen rights), which facilitates the distinction between formal citizenship and citizenship practices, and represents firm liberalization of the right of residence and work. At the same time, it ensures inclusiveness and expansiveness of the citizenship as collective emancipation, thus representing a kind of driving force which opens and broadens the boundaries of existing citizenship institutions. Citizenship is not only given from above, but is largely constituted from below, which empowers active citizenship. Therefore, the project of opening the borders is based on the basic participation of those who invoke *droit de cité*, even though they are formally not citizens of particular nation-state (Balibar 2004).

Modern citizenship contains two important deviations from the principles of universality, inclusiveness and democracy. The first deviation is the distinction between passive and active citizenship; between congenitally equal and inalienable rights on the one hand, and active participation in decision making on modalities of their implementation on the other. The second deviation is the relation between rights and nationality by birth. Therefore, the arrival of a foreigner or a refugee suspends the image of the national state as the one capable of granting human and citizen rights (Kurnik 2007).

Global citizenship is another alternative to nation-state citizenship. It is an even vaguer concept than European citizenship. Since global citizens (in contrast to European citizens as one of the post/transnational categories) are not a legal category, one can only imagine their existence and function as a kind of informal association of the like-minded. Because there is no such thing as global bureaucracy or global government that would protect global citizens, they can simply live and work within transnational frames, denying and avoiding national borders and national sovereignty. Van Steenbergen proposes more than just a technical definition of a global citizen, as he describes one in more holistic terms: individuals get to choose where they work and live and are no longer attached to (their) state or location where they were born (van Steenbergen 1994: 138). In an age when power was in the hands of ethnic and exclusivist models of nationalism, the political discourse of global citizenship was nevertheless on the rise (Carter 1997).

The concept of global citizenship originating in cosmopolitan values is based on early Western thought and was articulated during the Enlightenment. Interest in this concept has risen recently, since the world is facing increasing global integration, environmental interdependence and an information revolution. International organizations (mostly non-governmental) have brought about an understanding that individuals share a greater responsibility for tragedies, wars, pollution and other global problems, even when they do not happen in their immediate vicinity. The debate on global citizenship is at the same time an extension of the renewed debate about the role and power of citizenship within the state. In the eyes of many, the two concepts (global and national citizenship) exclude one another, because they supposedly address and defend contradicting positions (universal values and moral duties towards the whole of humanity versus values and duties towards a specific community, culture or state). From the perspective of nationalism and nation-state, cosmopolitanism seems to be shallow, since it lacks rooting in a specific and identifiable community. According to nationalism, cosmopolitanism addresses too abstract and unreal universal requirements. On the other hand, the cosmopolitan view perceives nationalism as narrow minded, exclusive, xenophobic and intolerant towards the rights of all who do not fall under the umbrella term of a national community (Carter 1997).

Although it seems at first glance that there is a conflict between nationalism and global citizenship, there is in fact not necessarily any clash between political affiliation to a specific country and the ideals of global citizenship, nor does a feeling of pride in a specific culture and identity exclude more universal feelings of global belonging. As Carter says in her article 'Nationalism and Global Citizenship', it is possible to find some forms of nationalism that are compatible with the original cosmopolitanism, especially those that express commitment to universal political principles, namely liberalism and republicanism, and take responsibility beyond national borders.

Despite the fact that there are concepts of and hopes to form a post-national citizenship that would be based solely on human rights, the national states remain the only entities that provide space in which human rights can be not only exercised but also advocated for all the excluded. Given that in everyday life the realization of human rights often coincides with citizenship status and the residence permit, it is of vital importance to defend the right to citizenship.

How is the above rather extensive discussion of citizenship relevant to our case? In Slovenia, the importance of citizenship status is shown in the case of the ex-Yugoslav nationals, and particularly the group of Erased. National citizenship was not granted to some groups of people since they were not perceived to be a part of the newly formed state. With Slovenia's entrance into the EU there were some hopes that European citizenship might be granted to all residents of the

Union. But since European citizenship was only an addition to the national one, the joining to the EU meant yet another exclusion for certain groups, since, from then on, they were deprived of two citizenships. A wider notion of citizenship, national, European or global, which would be based on the individual's residence or activity or other citizenship practices, would significantly contribute to solving the problem of stateless persons.

As Deželan demonstrates in his study, the Slovene language was a persistent issue during the nation-building process, so, according to this criteria, Slovene national continuity 'could only be traced through language, in a cultural sense, which in turn reinforced the dominant ethno-cultural conception of the nation' (Deželan 2011). In the end, the Slovene legislator adopted a 'milder' ethnic model in order to prevent the discontent of non-Slovene residents, but also to gain the approval of the international community for its respect of the democratic standards. This new citizenship regime, as Deželan argues, did not directly lead to widespread discrimination, but, nonetheless contributed to a situation where there were severe human rights violations (Deželan 2011).

Europeanization

In 1992, Slovenia became a member of the Organization for Security and Cooperation in Europe (OSCE) and a member state of the United Nations Organization. In 1993, Slovenia was accepted into the Council of Europe and subsequently applied for full EU membership in 1996. By joining these regional and international organizations, Slovenia bound itself to their laws and treaties concerning matters such as international relations, trading and human rights.

The key notion that the researchers of Slovene modern political history have been trying to determine for the last 20 years is the aspirations that have driven the Slovene independence movement. The dilemma is the following: first, was an independent state only a means of democratization that was not possible within the frame of Yugoslavia, or was democratization an 'excuse' to gain independence? And second, was the formation of an independent state a consequence of a wish to join the European Union, and democratization was the necessary condition to achieve this goal, or was democratization the elementary wish and goal that would automatically lead to Europeanization and membership?

The posed questions can be answered by strictly following the process of the disintegration of Yugoslavia and the decisions of the Slovene political (but also cultural and intellectual) elites, which influenced the course of Slovene independence. We could argue that in the late 1980s, the Socialist Yugoslav republic of Slovenia truly wished for the democratization of Yugoslavia. When the Yugoslav leaders showed ignorance towards these aspirations, the Slovene elite started a course of actions (free elections, new democratic legislation, civil rights movements and so on) to achieve it nonetheless. This eventually led to the formation of the independent state of Slovenia. But, at the crucial moment of the independence process, we can observe a rupture between the forces that focused on independence and state-building processes and forces whose primary goal was democratization. What they both had in common was the aspiration to join the European Community (and other relevant international associations) in the shortest possible time, since this represented a promise of economic development and a guardian of fundamental democratic standards. The EU was envisioned as a core political project, a cultural leap from something 'Balkan' and 'rotten' to something 'heavenly' to which the Slovene nation belonged. Therefore, like the process of independence, Europeanization was once again seen as a self-evident transition to a presumably higher level of democracy. Adhesion to the EU meant one final confirmation of Slovenia's European identity, and thus its most visible distinction from other former Yugoslav states. That became even more obvious with the adherence to the Schengen space, when, according to Vidmar-Horvat, a new era of negotiations on European identity and the identity of Europe began. Schengen Europe caused a redefinition of the hierarchies of inclusion and exclusion (Vidmar-Horvat 2009). During accession negotiations, Slovenia introduced several mechanisms vital to its democratic character. The European Commission pushed for a number of actions that led to some important reforms (on issues such as minorities, refugees and the Erased), so one can claim that the Slovene citizenship regime was significantly altered by accession to the EU.

The introduction of the new, European, citizenship, similar to the Yugoslav 'two-tiered' citizenship regime, meant new rights for workers, and a share of important citizenship rights for Slovene citizens within the EU, as well as for previously less protected groups and individuals. European citizenship also brought about voting rights for EU citizens in Slovenia, and for Slovene citizens in other EU states at local and European elections. This presented a major shift in the definition of a political community. Many studies of new democracies claim that the international context may influence transitions (toward democracy). As Rizman puts it, 'the Eastern and Central European countries had never before known more favourable conditions for change and, in terms of democratic consolidation, had never seen a combination of international factors more conductive to that process' (Rizman 2006: 151).

From the point of view of ethnic communities and the interests of the Member States, it is legitimate to use (ethno)national arguments when defending one's own strategic (national) interests, but one must, however, avoid isolationism and egoism. Like every other transitional society, Slovenia had similar problems in consolidating democracy. Or, as George Schöpflin says, 'all post-communist regimes are democratic in form but nationalistic in content' (Schöpflin cited in Rizman 1997: 225). Ethnically speaking, Slovenia as a rather homogeneous country has no problems with separatist, autonomist or even terrorist movements. However, this does not excuse it from the obligations it has towards its ethnic and other minorities, which want to preserve their distinctive ethnic (or some other form of) identity.

In the context of the EU, Slovenia has, as Rizman claims, a legitimate right to raise its voice when a critical issue affecting its interests is on the table. It may even exercise positive nationalism (similarly to Denmark) when trying to achieve a good platform for future negotiations and interests vital to its development. In this way one cannot doubt the positive value of nationalism, especially when seen through the lens of large nations that rarely face accusations of nationalism. That is, large nations often solidify their (sometimes even the most extreme) national interests by simply being formidable and with implicit threats, while small nations are easily labelled nationalistic and destructive (Rizman 1997: 233). It is clear that we have a moral duty to reject nationalism when it is directed against minorities or is a function of isolation and a closed society. We can, therefore, agree with liberal and defensive nationalism, but must reject and condemn totalitarian and fundamentalist nationalism.

The Extreme Right

Right-wing extremism as social phenomenon occurs in the ideological dimension, in political goals, political style, individual and socio-psychological outlooks, in socioeconomic stratification and finally, in all organizational characteristics in the function of filter. Political right-wing extremism is conditioned by intolerance, undemocratic principles, hatred and by focusing on 'extreme traditional' values (Ramet 1999: 150). All these elements are typical of any political right which equates the world with constant struggle against the 'hated others',[4] in the sense 'if you are not with us, you are against us' (Nastran-Ule 1999).

According to Rizman (1996: 20), the transition period is a complex and uncertain switch from a closed or semi-closed authoritarian society into an open democratic society. In such a time, every society is confronted with risk conditions. One of the negative effects of these risk conditions is the manifestation of a xenophobia that was suppressed under socialism. In Eastern and Central European post-socialist space, we can find similar politically organized expressions of such xenophobia and demagogic populism. As such, Craig Nation (Rizman 1998: 320) classifies Mečiar of Slovakia, Tudjman of Croatia, Janša of Slovenia, Berisha of Albania, Lukashenko of Belarus, and so on. All these politicians led political parties that can be described as 'Fürerparteien' in the West, for instance the party of Jörg Heider in Austria, Finni and Bossi in Italy and Le Pen's party in France (Rizman 1998: 259).

Kuzmanić claims that most post-socialist regimes 'needed' a certain amount of discrimination, nationalism, xenophobia, machismo, chauvinism, racism and other 'social and cultural' phenomena of European civilization to establish and strengthen their position and dominance (1994: 165–6). Klinar (1994) also searches for new manifestations of xenophobia in the uncertain transition period marked by

4 The concept 'hated Other' is taken from Freud, and is an idea of arbitrariness that allows violence against the target group (Ramet 1999: 3).

comprehensive restructuring, economic and social stagnation and, consequently, social exclusion. In the case of Slovenia, the foundations for xenophobia and intolerance were probably laid with mass internal migrations in the 1960s and 1970s, when the workforce started to migrate towards the west and the north of Yugoslavia, mostly to Slovenia. The beginning of modern Slovene xenophobia thus occurred with the interaction of Slovenes with the workforce from the south and with establishing contacts with their culture, habits and traditions (Kuzmanić 2003: 25).

In the late 1980s and the beginning of the 1990s, a new Slovene (ethnic and state) nationalism was born. One may see legitimate arguments and reasoning in its tendency for pluralization and democratization, and for greater autonomy within Yugoslavia. In that time, this kind of nationalist ideology was still rather benign, as it contributed to a consensus on key national interests and increased the willingness of people to carry the costs associated with pursuing these ambitions. With passing time, a new, negative nationalism prevailed due to the aforementioned reasons (bloody disintegration of the former state, socioeconomic crisis and the emergence of new nation-states). Its expression was intolerance towards groups and individuals originating in other republics of the former Yugoslavia (Klinar 1991).

In his debate on the radical right, Rizman points to the growing number of radical groups and organized right-wing parties and movements after Slovenian independence (Rizman 1998: 249–70). The reason for the late occurrence of the right-wing populism needs to be sought in the communist regime, which did not permit any political opponents and would suppress any form of opposition. In Slovenia, populists and right-wing groups draw upon the same principles as they do in other countries: prejudice and hatred towards immigrants, minorities, negative sentiments directed against neighbouring nations, and, in post-socialist states, also a continuing reference to communist failures and crimes. Extreme-right parties and their leaders manipulate the facts from the old, as well as the current systems, they make up phantom enemies, and their political stands change according to current political needs. In this regard one should and must mention also the role of the Roman Catholic Church, the ideas of which are close to the radical right-wing ones. One of the main features of the Slovene political right is the mass production of conflicts which cannot be resolved by the insufficiently developed democratic institutions (Rizman 2006: 133).

A common result of every nation formation is also the emphasized ethnic homogenization and nationalization of culture, together with attempts at quiet assimilation. Slovenia is no exception in this regard, and while forming a new nation-state, and especially in the first years after independence, this process allowed very limited ways of expressing non-Slovene ethnic identities (Kuzmanić 1999: 125).

Immigrants

In the decades after the Second World War, Slovenia was often a target destination for migrants from other republics of Yugoslavia. Silva Mežnarić (1982: 12) describes Slovenia as 'one of the most concentrated immigration areas in Europe'. Official statistics show that between 1960 and 1980 more than 115,000 people from other Yugoslav republics moved to Slovenia. Ethnic composition changed significantly. While at the end of the Second World War more than 90 per cent of the population in Slovenia were ethnic Slovenes, the demographic picture after 1980 demonstrates that the percentage of those who identify as Slovenes reduced to 80 per cent. According to statistical data from 1978, approximately 18,000 Bosnians moved to Slovenia between 1961 and 1971. In the period from 1971 to 1975, 24 per cent of all immigrants to Slovenia came from Bosnia and this share grew to 46 per cent in the years from 1976 to 1981. With wars in Croatia, and later in Bosnia and Herzegovina, new waves of migrants reached Slovenia (this time refugees). The total number of refugees exceeded 45,000 people, most of whom later moved to Western Europe, Canada and the USA. In May 2001, 2,646 of them were still living in Slovenia (Official census from 2001). This was also a period of extensive media discourse that created additional fear and distance. A study shows that in just 35 days, 162 articles were published, most of them using terms such as 'floods of people' and 'massive waves of refugees' (Jalušič 2003: 16). Slovenia, once ethnically rather homogeneous, started its independence in a different, more ethnically plural state. It seems no one was prepared to acknowledge this new condition as we may observe from the political and public discourse on the topic.

Ethnic groups of Serbs, Croats, Bosnians, Macedonians, Montenegrins and Kosovo Albanians became de facto ethnic minorities after the independence of Slovenia, even if the Slovene state did not recognize them as such (Klopčič, Komac and Kržišnik-Bukić 2003, Komac and Medvešek 2005). Up until then those communities were formed around their religious and cultural activities, never addressing any political demands. Regardless of their current political passiveness and the status Slovene politics is willing to give to these 'new minorities', Slovenia will have to rethink its integration policies – not only because of the above-mentioned groups of first, second and even third generations of immigrants from former Yugoslavia, but also in terms of new migrations from other EU Member States, immigrants from Africa, Asia, South America and elsewhere.

The Erased

Among the above described groups of Yugoslavs who moved to Slovenia prior to its independence there is also a group of about 25,000 people who were erased from the Registry of Permanent Residents of Slovenia in 1992. They had registered their permanent residence in Slovenia but did not acquire citizenship of the Republic of Slovenia after its proclamation of independence. Thus, on 26 February 1992 they were transferred under the jurisdiction of the *Aliens Act*, which led to the deprivation of their former permanent resident status. They became foreigners, living in Slovenia illegally. This way, they were left without political, economic and social rights. The Slovene authorities informed those residents who did not automatically become part of the newly formed civic body that they must apply for the new Slovene citizenship if they wished to acquire it, but they did not inform these residents that failure to obtain Slovene citizenship would result in the cancellation of permanent residence status as well. In fact, the *Aliens Act* had not originally predicted such consequences, but since it provided the former Yugoslav nationals neither with a transitory status nor with the automatic acquisition of an alien status, the Ministry of Internal Affairs *solved* the legal gap with a decree saying that the records concerned should be transferred from the register of permanent residents to the aliens register.

When analysing the erasure phenomenon, the question that arises is how did the erasure occur and, more importantly, why was it possible? One could argue that the legal gap in the *Aliens Act* actually allowed its occurrence. Since that gap had to be fixed, the Ministry of Interior executed a discriminatory decree. But the complexity of the phenomenon of the erasure cannot be satisfactorily explained solely from a legal point of view.

The social atmosphere during the independence process was instrumental to the erasure. The newly established state, as demonstrated above, wanted to distinguish itself from its Balkan vicinity as radically as possible, which affected the process of formation of the new Slovene citizenship. In order to achieve national homogeneity, the newly established state nationalized the right to citizenship, and thus put all ethnically non-Slovene residents into an inferior position. In this context, the Erased were easily labelled as the *cultural Other*, unsuitable members of the new state. Their non-acquisition of Slovene citizenship was understood as their lack of faith in the new state; they were therefore labelled as traitors, aggressors and opponents of Slovenia. They remained the representatives of the Balkans, which resulted in the explicit racist discourse of binary *us* vs. *them*. As Vlasta Jalušič observes, various social spheres contributed ideologically to the formation and conservation of the racist discourse with their indifference to discrimination. Although numerous officials encountered many erased persons, instead of notifying their superiors about the illegalities, they merely followed the discriminatory rules (Jalušič 2007). What is more, prior to erasure, the Erased had friends, acquaintances, neighbours and colleagues in Slovenia, who, after the erasure, mostly became indifferent to the discrimination and thus helped to maintain the illegal status of the Erased unchanged. Furthermore, the silence of the media played a significant role in keeping the Erased in their marginal position (Lipovec-Čebron 2007). Society's prevailing perception was that the erasure had been but a consequence of the normal functioning of the state, not its deviation, and especially that it had not been intentional or systematic. Jalušič names this phenomenon *organized innocence*, which emerged as an attempt to avoid responsibility and to make it collective (Jalušič 2003).

It was only in 1999 that the unconstitutionality of the erasure was legally recognized by the constitutional court. In 2003, in its last judgement on the erasure, the constitutional court imposed on the Ministry to issue supplementary decisions for those Erased who had by that time managed to obtain legal status, thus re-establishing their permanent residence in Slovenia as of 26 February 1992. It was only with the re-election of a centre-left government in 2008 that the constitutional court judgement from 2003 was implemented. For those whose legal status had not yet been set, the Ministry adopted a special act, which presented most of the Erased with an opportunity to regain their permanent resident status. The adopted act is completely in line with the constitutional court decision and therefore some groups of erased persons are excluded. The act is based on individual treatment and it expects the Erased to initiate the proceedings, even though they were not responsible for the erasure. Furthermore, the act does not regulate the indemnity questions nor does it ascertain responsibility for the erasure (Kogovšek 2007).

The Erased have also spread awareness of the problem to some European institutions, thus taking advantage of the new source of political pressure on the Slovene authorities. They succeeded with their lawsuit at the European Court of Human Rights, which obliged Slovenia to grant the Erased their human rights. Whether (or not) Slovenia is to abide by the judgement of the court's decision is set to become an indicator of the democratization and Europeanization of Slovenia and a task for the EU to overcome its double standards when defending human rights. Since 'the rather hidden fate of the Erased failed to make a substantial dent in the overall success rate of Slovenia, which has been widely seen in international spheres as the only success story of the dissolution of Yugoslavia and a deserving member of the "2004 team" – i.e. the first post-1989 enlargement of the EU towards Central and Eastern Europe' (Shaw 2010: 22), acceptance of the European Court's decision will put Slovenia's democratic standards and European values to the test.

The Constitutional Court of the Republic of Slovenia has twice established that an unconstitutional situation was involved, namely that the erasing was illegal and that the government failed to address its consequences properly. The unconstitutionality and illegality of the erasure has been noted by the Constitutional Court indirectly also in seven other cases. Slovenia put right the injustice brought upon the Erased and closed a painful chapter in the history of its young statehood only in 2010, 18 years after the erasure, 11 years since the first and seven since the second decision of the Constitutional Court.

The Roma

In general, sociological research on the Roma usually focuses on the connection their low social status has to the status of highly disadvantaged ethnic minority. Roma are ranked among 'lower social groups', affected by poverty and placed at the very bottom of the ethnic stratification structure. Roma poverty is to be associated with its congenital ethnic status (Klinar 1991: 24). The social status of the Roma is ranked lower than the social status of the majority of the population as well as that of other ethnic minorities, which puts them among the most vulnerable marginalized groups. These findings apply both to general social status, as well as the special social statuses, such as material, educational and professional. Obviously, due to their exclusion from economic and political institutions, they do not possess any socially powerful means, and have almost no opportunities on the labour market or in the political sphere. Roma are subjected to discrimination, segregation, ethnic distance, social disorganization and pathology.

In Slovenia, the situation of the Roma is not much different from that in other post-socialist countries. The status ascribed to Roma from the point of the majority society is that of 'eternal foreigners'. Their situation is affected by the local communities, which rejected the integration of Roma into their political structures, as well as by the academic differentiation between autochthonous and non-autochthonous Roma. Such perceptions even contributed to demands to relocate some of the Roma groups without their consent.

In Slovenia, Roma communities live in two parts of the country: in the northeast (the Prekmurje region) and the southeast (the Dolenjska region). Approximately 70 per cent of Roma children attend school in the Prekmurje region, compared to 39 per cent in the Dolenjska region. The most common constraints that Roma have to face include the lack of financial resources (high unemployment), unheated houses and inadequate living conditions in general (many of their settlements are illegal). However, Roma in Slovenia enjoy specific judicial protection under the 1993 Law on Local Self-government. This means that Roma must have at least one representative in the municipal council (imposed in communities with a certain percentage of Roma, which is the case in 20 local communities). On 31 March 2007, the Slovene parliament passed a new Law on Roma community (required and foreseen by the Slovenian Constitution in 1991), making Slovenia the only country in the EU with special legislation for the protection of the Roma ethnic minority. Regardless of this new political arrangement, Roma still face many problems, and Slovenia will have to make a shift in political and administrative culture to establish their equal treatment.

Muslims

In Slovenia, according to the 2002 census, approximately 50,000 citizens (2.7 per cent of the whole population) identified as Muslims (Statistical Office of the Republic of Slovenia 2002). They are predominately members of the Bosnian ethnic community who emigrated to Slovenia between 1960 and 1991, including their descendants. The highest representative body of Slovene Muslims is the Meshihat of the Islamic Community in Slovenia, an institution that falls within the Meshihat of the Islamic Community in Bosnia and Herzegovina. Osman Djogić, a former leader of Slovene Muslims, established a new organization, the Slovene Muslim Community, as a result of serious disagreements on how autonomous/independent from religious authorities in Bosnia and Herzegovina the Slovene community should be.

The situation and the status of Slovene Muslims are best described by outlining their decades-long desire to build an Islamic religious and cultural centre in Ljubljana. For many years, the local community has tried to prevent such plans, while enjoying both the explicit and the implicit support of the Roman Catholic Church and right-wing political parties. The ongoing story of mosque construction in Ljubljana, as well as the intolerant public discourse is an evident case of historically rooted Islamophobia in Slovenia. Dragoš (2004: 5–12) classifies this issue within three time periods, characterizing them as a transformation from latent xenophobia to explicit xenophobia, with manifestations of institutionalized Islamophobia. The first period is marked by the problems the Muslim community had to face when it first applied for mosque construction (1969) up until the first official rejection from the authorities (2001). During the whole of this time, the local authorities avoided making a decision, claiming that there was no adequate location for 'this kind' of religious building. The second period saw even greater and more explicit articulation of negation and opposition. The third period, from 2003 to 2007, is characterized by Dragoš as a dangerous response, a manifestation of xenophobia, institutionalized by an initiative for a national referendum (Dragoš 2004: 13).

There are many irrational arguments against the construction of a mosque in Ljubljana. Dragoš (2003: 48–53) claims that these arguments are produced by politicians and generated by the mass media. They are being formulated mainly by the right-wing political circles, driven by the opportunistic attitude of gaining political support. The 30 May 2007, however, was the turning point in the whole process. The Mayor of Ljubljana, Zoran Janković, and the mufti of the Islamic community in Slovenia, Nedžad Grabus, signed the letter of intent regarding the sale of land needed to build the Islamic cultural and religious centre. After being moved from one peripheral part of the capital to another, its final location places the religious and cultural complex closer to the city centre. The strongest argument against the mosque was of an 'architectural' nature, namely the height of the minaret. In a relatively fierce and fiery political campaign, the initiators opted for a low minaret that would not 'destroy' Ljubljana's urban (Catholic) skyline. They even promoted the idea of a national referendum, but in the end the municipality managed to prevent the referendum, so the original architectural plan that envisioned a 40-metre high minaret remains.

Yugosphere and Yugonostalgia

In this section, two relatively new phenomena, namely Yugosphere and Yugonostalgia, will be addressed. Very few authors have given a thought to the kinds of connections and contacts that have been established throughout post-Yugoslav spaces, especially considering the fact that we now speak of very different societies, for which it is almost impossible to find one common definition.

The Yugosphere is a term used to describe the present-day areas which once formed Yugoslavia. As a result not only of the similarity of their languages, but also of their long common history, we witness numerous ties amongst people, even though the newly formed states keep favouring differences over commonalities. The memory of the Yugoslav period is most often related to the socialist system and its social security, but also to a time of economic stability and the achievements of the former common state.

The National Public Radio journalist Amy Standen (2006) reported in '"Yugonostalgia" Takes Hold in Slovenia' that in spite of the good economic situation, many people looked back longingly, remembering the days when Slovenia was still part of Yugoslavia. Not only a number members of the older generations, but also many young people who have no first-hand experiences would decorate their college dorm rooms with posters of President Tito. Crowds provide a lively response to 'Računajte na nas'/'Rely on us', a song written in 1978 as an anthem to socialism and to Josip Broz Tito. Many would agree with the interviewee in the article, who claimed that life felt easier when Slovenia was known as Yugoslavia and when people had more secure jobs, when social security was at a very high level, when pensions were guaranteed. 'You didn't get much, but with that you could live very well', claimed the owner of a bar in Ljubljana. Most of the younger generations have no political reservations towards the post-Yugoslav space so they are likely to travel to Sarajevo, Skopje or Belgrade as tourists, knowing of past events but having no direct memories of wars, refugees and the breakup of Yugoslavia.

Similarly in 2008, *The Independent* noted a surprising number of former citizens from more or less all the former republics and described how passionately devoted some people were in regard to memories of the former country and, in particular, its President Tito. The author Peter Popham reported on the remarkable rise of Yugonostalgia, and emphasized Slovenia as a bastion of sentiment for 'the good old days'. It seemed strange to him that the one republic that 'got out of the federal state without suffering any ill consequences and is now safely installed in the EU' was now claustrophobic because of its size, and thinks of Yugoslavia where it felt safer and more influential. He also wrote about a group of 50 young Slovenes who visited Belgrade to celebrate Tito's 116th birthday, accompanied by a Tito impersonator and children dressed up as Partisans (Popham 2008).

One article talks about this new and supposedly widespread phenomenon as of Yugoslavia 'patching itself together' (Spaskovska 2009). According to the author, the supra-ethnic trans-Balkan cultural inclination slowly turns, on one hand, into a more conscious realization of the unsustainability of the parochial/tribal self-isolation from the 1990s, and on the other, into a more emotionally and culturally framed awareness related to the vibrant interconnected cultural life in this region. The author claims that the economic ties and the more pragmatic market-oriented mind-set is what drives the region towards a greater integration. Yugoslavia, with its population of 22 million, was a relatively big market, particularly when compared to the post-1992 fragmented political and cultural spaces. Referring to Sabrina Ramet's 1994 book, *Rocking the State: Rock Music and Politics in Eastern Europe and Russia*, it is the emerging shared cultural sphere which is likely to boost the figures economists are concerned about: 'Increased mobility for tourism and educational purposes as well as music, film, art and literature make the physical borders in this region more porous and communication more effective', she claims while also mentioning some commercial projects such as MTV's branch – MTV Adria – covering all former Yugoslav republics (Ramet 1994).

We can identify both positive and negative senses of Yugonostalgia when referring to Slovenia (the same probably applies also to other ex-Yugoslav states). While with its positive connotations this kind of attitude and sentiment refer to an emotional attachment to the idealized and desirable aspects of life in Yugoslavia (economic security, multiculturalism, non-alignment movement, to name a few), compared to the unstable situation in economic and political transition, the negative sense has been used by nationalists whose objective is to portray their critics as anachronistic, unrealistic and unpatriotic. In Slovenia, the term Yugonostalgic was often replaced by pejorative labelling of 'balkanophiles' or 'udba-communists', and, according to Dubravka Ugrešić, the term Yugonostalgic is also used to discredit a person as a 'public enemy' and a 'traitor' (1998: 231).

In general terms, and as Svetlana Boym, probably one of the most prominent authors on nostalgia (both theoretically and in post-socialist countries), puts it, we may divide the concept of nostalgia into two categories, restorative and reflective. While restorative nostalgia focuses on 'returning home' in an attempt to restore the past and the way it was before, reflective nostalgia is concerned with pain and longing. In this case there is no attempt to restore the past but it is instead somewhat of a dwelling on the feeling of remembrance (Boym 2001). Velikonja, who is specifically focusing on the concept of Yugonostalgia, divides nostalgia even further into feeling/idea nostalgia and materialized nostalgia. The first concept is nostalgia that is present in individuals and in their collective and individual memories of the past, while the latter is present through material objects such as buildings, monuments and souvenirs. We can find wide range with Yugoslavia- and Tito-related items sold and exchanged on the internet, in antiquities markets in Ljubljana and other sites (partisan hats, five-pointed stars, Tito images, *Yugo-music*), mostly to fulfil the nostalgia for socialist Yugoslav times. As Velikonja points out in his *Titostalgia*, these individuals, involved in fascination for Yugoslavia, do not necessarily have to have lived through or experienced a specific time period in order to be nostalgic for it; in many cases it is simply an act of protest and rebellion against the current political and economic situation (Velikonja 2008: 30).

Conclusion

Slovenia escaped Yugoslavia rather successfully, compared to the war zones in Croatia, Bosnia and Herzegovina or Kosovo. Its rapid development and international successes (joining major international organizations such as the EU, NATO, UN, OECD) distanced it from the Balkans for many years, but at the same time, people kept their previous intimate ties (private, family) and other relations (former business partners, economic networks of some companies).

Sentiments of nostalgia started immediately after the new country was born, and grew even more intense with time. Slovene tourists never truly left their traditional destinations in Croatia and Montenegro, and now the younger and unburdened generations keep rediscovering new destinations (like Belgrade, Sarajevo and Skopje). Bosnia and Herzegovina has always been an interesting destination for Slovene business as well as individuals, culturally fascinated by Sarajevo, the ancient town of Mostar. In recent years we may observe the deepening and intensifying of political contacts at the highest levels; Slovenia was very active in peacekeeping operations in the Balkans (first Bosnia and Macedonia and now Kosovo). The Slovene economy has always perceived the Balkan market as its 'own' (of all Slovene foreign direct investments, 67 per cent are in the western Balkans). The economic presence of Slovenia in the ex-Yugoslav republics has, on the other hand, never translated into political influence, partly due to its size, but also because of all the resentment between the political and cultural elites.

With all due seriousness and respect for the Slovenian accomplishments of the last 20 years, we may state that Yugoslavia was a successful story of a historical attempt at unification of the southern Slavic peoples into a common political frame. The complex circumstances (international, ethnic, political, cultural, historical, religious, ideological and other) affected the attempt to create, and then to disintegrate, a second Yugoslavia. The author of this chapter neither hopes nor expects to witness the formation of a new Yugoslavia in the near or even more distant future. Nevertheless, we may expect and hope that now, after having seen too much 'history', too much suffering, hatred, and lost generations in too short a time, the small states which once formed a country named Yugoslavia will finally establish partner relations, seeking joint projects and common ways, building on dialogue and cooperation. The European Union, notwithstanding its current difficulties, is still the only true opportunity for the post-Yugoslav region. Our common future is imminent, and the new, unburdened generations are a proof that such a future will also be creative and peaceful.

References

Alvis, R.E. 2005. *Religion and the Rise of Nationalism: A Profile of an East-Central European City*. Syracuse, NY: Syracuse University Press.

Arendt, H. 2004. *The Origins of Totalitarianism*. New York, NY: Schocken.

Balibar, E. 1994. Predlogi v zvezi z državljanstvom. *Časopis za kritiko znanosti*, 22(172–3), 45–54.

Balibar, E. 2004. *We, the People of Europe? Reflections on Transnational Citizenship*. Princeton, NJ: Princeton University Press.

Bebler, A. 2010. The Dissolution of the Yugoslav Federation: Its Causes and Consequences. *Europejski prezegląd prawa i stosunków międzynarodowych*, Issue 1, 31–52.

Borak, N. 2000. Economic Background to National Conflicts in Yugoslavia, in *Economic Change and the National Question in the Twentieth-Century Europe*, edited by A. Teichova, H. Matis and J. Pátek. Cambridge: Cambridge University Press, 310–37.

Boym, S. 2001. *The Future of Nostalgia*. New York, NY: Basic Books.

Carter, A. 1997. Nationalism and Global Citizenship. *The Australian Journal of Politics and History*, 43(1), 67–81.

Connor, W. 1994. *Ethnonationalism: The Quest for Understanding*. Princeton, NJ: Princeton University Press.

Conversi, D. 2003. Conceptualizing Nationalism: An Introduction to Walker Connor's Work, in *Ethnonationalism in the Contemporary World: Walker Connor and the Study of Nationalism*, edited by D. Conversi. London: Routledge, 1–23.

Dedić, J. 2003. Diskriminacija v postopkih pridobivanja slovenskega državljanstva, in *Izbrisani: organizirana nedolžnost in politike izključevanja*, edited by J. Dedić. Ljubljana: Mirovni inštitut, 23–85.

Deželan, T. 2011. The Janus Face of Slovene Citizenship. *Citizenship in Southeast Europe Project*. Available at: http://www.citsee.eu/citsee-study/janus-face-Slovene-citizenship [accessed: 20 March 2012].

Dragoš, S. 2003. Katoliška identiteta Slovencev? *Annales: Anali za istrske in mediteranske študije*, 13(1), 39–54.

Dragoš, S. 2004. Islamofobija na Slovenskem. *Poročilo Skupine za spremljanje nestrpnosti*, Issue 3, 10–27.

Fink-Hafner, D. 1995. The Disintegration of Yugoslavia. *Canadian Slavonic Papers*, 37(3–4), 339–65.

Gellner, E. 1964. *Thought and Change*. Chicago, IL: University of Chicago Press.

Glenny, M. 1996. *The Fall of Yugoslavia: The Third Balkan War*. London: Penguin Books.

Glover, J. 1997. Nations, Identity, and Conflict, in *The Morality of Nationalism*, edited by R. McKim and J. McMahan. Oxford: Oxford University Press, 11–30.

Horn, F. 1998. Conception and Principles of Citizenship in Modern Western Democracies, in *Citizenship and State Succession: Proceedings of the UniDem Seminar Organized in Vilnius, 16–17 May 1997*. Vilnius: University of Vilnius Press, 39–87.

Jalušič, V. 2003. Organizirana nedolžnost, in *Izbrisani: organizirana nedolžnost in politike izključevanja*, edited by J. Dedić. Ljubljana: Mirovni inštitut, 7–23.

Jalušič, V. 2007. Organizirana nedolžnost in izključevanje: nacionalne države in državljanstvo po vojni in kolektivnih zločinih. *Časopis za kritiko znanosti*, 35(228), 101–21.

Janjić, D. 1992. *Disintegration of Yugoslavia and Inter-Ethnic Conflicts. Nation and State: Small Nations and Ethnic Minorities in the Emerging Europe*. International Scientific Conference on Minorities for Europe of tomorrow. Maribor, 3–5 February 1992.

Klinar, P. 1991. Romi med revščino in etnično deprivilegiranostjo. *Razprave in gradivo*, Issue 25. Ljubljana: Inštitut za narodnostna vprašanja.

Klinar, P. 1994. O nacionalni identiteti in etnonacionalizmih: Slovenija v prehodnem obdobju. *Teorija in praksa*, 31(5–6), 421–37.

Klopčič, V., Komac, M. and Kržišnik-Bukić, V. 2003. *Albanci, Bošnjaki, Črnogorci, Hrvati,Makedonci in Srbi v Republiki Sloveniji: Položaj in status pripadnikov narodov nekdanje Jugoslavije v Republiki Sloveniji*. Ljubljana: Inštitut za narodnostna vprašanja.

Kogovšek, N. 2007. Izbrisani: predlog ustavnega zakona kot negacija pravne države. *Časopis za kritiko znanosti*, 35(228), 177–93.

Komac, M. and Medvešek, M. 2005. *Percepcije slovenske integracijske politike: Zaključno poročilo*. Ljubljana: Inštitut za narodnostna vprašanja.

Kralj, A. 2009. *When Ć Becomes Č: Discrimination of Unrecognized National Minorities in Slovenia*. Ljubljana: University of Primorska Science and Research.

Kučan, M. 1992. Speech at the General Assembly of the UN – Acceptance of New Members. New York City, 22 May 1992.

Kurnik, A. 2007. Izbrisani gredo v nebesa. *Časopis za kritiko znanosti*, 35(228), 122–31.

Kuzmanić, T. 1994. Postsocializem in toleranca ali Toleranca je toleranca tistih, ki tolerirajo – ali pa ne! *Časopis za kritiko znanosti*, 22(164–5), 165–83.

Kuzmanić, T. 1999. *Bitja s pol strešice: slovenski rasizem, šovinizem in seksizem*. Ljubljana: Open Society Institute.

Kuzmanić, T. 2003. Ksenofobija v nekdanji SFRJ in v postsocialistični Sloveniji, in *Poročilo skupine za spremljanje nestrpnosti 02*, edited by K. Roman and T. Trplan. Ljubljana: Peace Institute, 14–34.

Kymlicka, W. 2007. *Multicultural Odysseys: Navigating the New International Politics of Diversity*. Oxford: Oxford University Press.

Lichtenberg, J.1997. Nationalism, For and (Mainly) Against, in *The Morality of Nationalism*, edited by R. McKim and J. McMahan, New York, NY: Oxford University Press, 158–76.

Lipovec-Čebron, U. 2007. Metastaze izbrisa. *Časopis za kritiko znanosti*, 35(228), 59–75.

Mandelc, D. and Učakar, T. 2011. Perforated Democracy: Disintegration, State-Building, Europeanization and the Erased of Slovenia. *Revija za sociologiju*, 41(1), 27–51.

Matić, D. 2006. Je li nacionalizam stvarno toliko loš: Slučaj Hrvatske, in *Demokratska tranzicija u Hrvatskoj: Transformacija vrijednosti, obrazovanje, mediji*, edited by S. Ramet and D. Matić. Zagreb: Alinea, 261–84.

Mežnarić, S. 1982. *Delavci iz drugih republik in pokrajin v Sloveniji*. Ljubljana: RC za samoupravljanje RS ZSS.

Nastran-Ule, M. 1999. Social Psychology of Prejudice, in *Prejudices and Discrimination: Selected Socio-Psychological Studies*, edited by M. Nastran-Ule. Ljubljana: Znanstveno in publicistično središče, 301–12.

Özkirimli, U. 2000. *Theories of Nationalism: A Critical Introduction*. London: Palgrave.

Petrović, T. 2009. *A Long Way Home: Representations of the Western Balkans in Political and Media Discourses*. Ljubljana: Peace Institute.

Popham, P. 2008. Bringing back Tito. *The Independent*, 1 March 2008.

Ramet, S.P. 1994. *Rocking the State: Rock Music and Politics in Eastern Europe and Russia*. Boulder, CO: Westview Press.

Ramet, S.P. 1996. *Balkan Babel: The Disintegration of Yugoslavia from the Death of Tito to Ethnic War*. 2nd edition. Boulder, CO: Westview Press.

Ramet, S.P. 1999. Defining the Radical Right: Values and Behaviours of Organized Intolerance in Post-Communist Central and Eastern Europe, in *The Radical Right in Central and Eastern Europe since 1989*, edited by S.P. Ramet. University Park, PA: Pennsylvania State University Press, 1–27.

Repe, B. 2001. *Slovenci v osemdesetih letih*. Ljubljana: Zveza zgodovinskih društev Slovenije.

Rizman, R. 1996. The Radical Right Politics in Slovenia, in *Europe and the Balkans: How They View Each Other. Identities and Perceptions, Autarchies and Differences the End of the 20th Century*, edited by S. Bianchini. Ravenna: Longo Editore, 1–38.

Rizman, R. 1997. *Izzivi odprte družbe: Sociološki komentarji (1989–1996)*. Ljubljana: Liberalna akademija.

Rizman, R. 1998. The Radical Right Politics in Slovenia, in *State Building in the Balkans: Dilemmas on the Eve of the 21st Century*, edited by S. Bianchini and G. Schöpflin. Ravenna: Longo Editore, 315–37.

Rizman, R. 2006. *Uncertain Path: Democratic Transition and Consolidation in Slovenia*. College Station, TX: Texas A&M University Press.

Shaw, J. and Štiks, I. 2010. The Europeanisations of Citizenship in the Successor States of the Former Yugoslavia: An Introduction. *Citizenship in Southeast Europe Project. Working paper series 2010/01*. Edinburgh: University of Edinburgh, School of Law.

Smith, A.D. 1981. *The Ethnic Revival*. Cambridge: Cambridge University Press.

Smith, A.D. 1991. *National Identity*. Las Vegas, NV: University of Nevada Press.

Smith, A.D. 1995. *The Ethnic Origins of Nations*. Oxford: Blackwell.

Smith, A.D. 2001. *Nationalism: Theory, Ideology, History*. Cambridge: Polity Press.

Spaskovska, Lj. 2009. The 'Yugosphere'. 28 September 2009. Available at: http://www.law.ed.ac.uk/ecclblog/blogentry.aspx?blogentryref=7915 [accessed: 20 March 2012].

Standen, A. 2006. 'Yugonostalgia' Takes Hold in Slovenia. *National Public Radio*, 9 October 2006.

Statistical Office of the Republic of Slovenia 2002. *Census 2002*. Available at: http://www.stat.si/popis2002/si/rezultati/rezultati_red.asp?ter=SLO&st=8 [accessed: 10 January 2012].

Tomc, G. 1993. Slovenski etnični značaj. *Nova revija*, No. 134–35, Summer 1993.

Tuccillo Castaldo, I. 2007. Imobilizirana državljanstva. *Časopis za kritiko znanosti*, 35(228), 139–47.

Ugrešić, D. 1998. *The Culture of Lies: Antipolitical Essays*. University Park, PA: Pennsylvania State University Press.

van Steenbergen, B. 1994. *The Condition of Citizenship*. London: Sage.

Velikonja, M. 2008. *Titostalgia: A Study of Nostalgia for Josip Broz*. Ljubljana: Peace Institute.

Vidmar-Horvat, K. 2009. *Zemljevidi vmesnosti: Eseji o evropski kulturi in identiteti po koncu hladne vojne*. Ljubljana: Sophia.

World Bank 1991. *The Challenge of Development*. New York, NY: Oxford University Press.

Chapter 3
Expanded Accession Conditionality: The European Union's Leverage on Refugee Return to Croatia

Antonija Petričušić and Ljubomir Mikić

The respect for and protection of minorities that were made part of the EU Copenhagen accession criteria adopted in 1993 remain binding criteria for all countries aspiring to EU membership now or in the future, including the present Western Balkan (potential) candidates. This chapter argues that an expanded element of the minority criterion, namely, the necessity to allow for return of refugees, has been attached to the pre-accession minority criterion in Croatia and several other (potential) Western Balkan candidates. Accordingly, by insisting upon refugee return and adequate responses to the manifold consequences of the inter-ethnic conflicts of the 1990s, the EU is contributing not only to a proper accommodation of relevant issues concerning national minorities and their protection in Croatia, but also to sustainable peace in the entire region, and beyond. This expanded, novel element of conditionality speaks in favour of the EU as a motivator of domestic changes in the sphere of fundamental rights. Subsequently, the chapter provides an overview of the requirements Croatia has been faced with in the field of refugee return in the course of the EU accession process, arguing that without the EU's strict and coherent conditionality on refugee return, the – already limited – success of refugee return would have been much poorer.

Introduction

As a region of relatively recent inter-ethnic conflicts, the Western Balkans are burdened with an abundant number of displaced persons who fled from the place of their traditional residence due to threats of persecution and/or lack of effective protection by the authorities. It is estimated that the violent breakup of the former Yugoslavia resulted in more than three million displaced persons, causing the largest refugee crisis in Europe since the Second World War. A significant number of displaced persons sought refuge in more than a hundred countries. A number of them were resettled to 'third countries' through programmes arranged by various international organizations. In 2008 it was estimated that nearly half a million persons, mostly of minority ethnic origin, lived in the region as refugees and internally displaced persons (IDPs) (UNHCR 2008). More than 15 years since the end of the conflicts, it is estimated that there are still about 438,000 refugees and other displaced persons (Council of Europe 2012: 5). Therefore, refugee return in the Western Balkans can, at best, be considered as only a partial success and it is 'unlikely to stem the tide of ethnic homogenization and territorialisation' (Bieber 2007: 26) that is the Western Balkan reality.

Compare the time elapsed since the war era to the track record of refugee return: it is no daring conclusion that complete refugee return is never going to happen. Nevertheless, the – at least declaratory – foreign policies of the Western Balkan countries articulated in the European Union membership aspiration has stimulated, to a certain extent, cross-border refugee return as well as repossession of pre-conflict property. Pre-accession criteria that require respect for and protection of minorities, introduced in the Copenhagen pre-accession criteria designed in 1993, remain binding for the Western Balkan (potential) candidate countries. However, in the case of the Western Balkan countries, the conditions defined by the Stabilisation and Association Process are also a fundamental element of the accession policy, requiring, in addition, cooperation with the UN International Criminal Tribunal for the former Yugoslavia (ICTY), regional cooperation, prosecution of war crimes and refugee return (European Commission 2005: 8, European Commission 2008: 5).

The requirement of refugee return, attached to the political set of the pre-accession criteria, set by the European Union for Croatia (and other Western Balkans states), seems to expand a normative requirement vis-à-vis the refugee protection set forth in relevant UN instruments. Namely, under the auspices of the UN, two international law instruments were developed to ensure the adequate treatment of refugees and protect their human rights: the 1951 Convention Relating to the Status of Refugee (hereinafter: Refugee Convention) and its 1967 Protocol (Damjanović and Tomičević 2004). The Refugee Convention lays down and promotes the basic human rights of refugees. It defines the term 'refugee' and their rights, as well as the legal obligations of states. It moreover articulates the important international standard of *non-refoulement*: the guarantee not to be returned involuntarily to a country where a refugee might face persecution. It helps them to repatriate to their homeland when conditions permit, to integrate into states of asylum or to resettle in third countries.

The Refugee Convention defines a refugee as 'a person who is outside his or her country of nationality or habitual residence; has a well-founded fear of being persecuted because of his or her race, religion, nationality, membership of a particular social group or political opinion; and is unable or unwilling to avail him or herself of the protection of that country, or to return there, for fear of persecution' (United Nations 1951: Article 1A(2)). Obviously, as they are not protected by their own governments, protection of refugees is ensured by international law that sets forth protection duties for those states under which territories they search refuge. The principle of *non-refoulement* contained in the Refugee Convention states that refugees shall not be returned to a country where they face serious threats to their lives or freedom (United Nations 1951: Article 33). Since this principle is considered a rule of customary international law, it is legally binding and must be applied regardless of whether a country ratified the Refugee Convention or its 1967 Protocol. Once the basis for refugee status ceases to exist, the status expires (for example, in case of voluntary repatriation to home countries once the situation there permits such return or if refugees integrate or become naturalized in their host countries and stay permanently).

In order to bring the refugee-return process to an end, refugees and displaced persons have several options at their disposal, each requiring clear political will on the part of the states concerned and their financial support. The primary state duty and responsibility is to allow for voluntary return of refugees and displaced persons to their homes or places of habitual residence in safety and dignity. However, refugees and displaced persons might opt not to return. In that case, states can either financially back their voluntary resettlement or support local integration of refugees into the communities in which they are presently located. In order for those processes to be satisfactory, returnees' 'decisions should be based on updated, reliable data on the number and situation of forcibly displaced persons and returnees' (Council of Europe 2012). Apart from establishing favourable return conditions and providing the respective necessary means, the authorities of the Western Balkan countries also need 'to ensure the full participation of minority members in all relevant action-planning and decision-making' (Council of Europe 2012).

By applying and scrutinizing the rational choice theory paradigm, this chapter will therefore argue that the EU accession process has played a role in intensifying the efforts of the countries' leadership in finding solutions for closing the refugee chapter, and describe the gradual changes in the Croatian policy towards refugee return. It is possible to analyse the EU accession process of the Western Balkans through different social theories. It appears that the rational choice approach is the most appropriate theory to explain the change of domestic elites towards a rectification of the consequences of wars fought throughout the 1990s. Rational choice theory is a broad theoretical approach, often used by social scientists to explain 'the nature of human agency and its relationship to social structures, the role of ideas and material forces in social life, the proper form of social explanations and so on' (Wendt 1999). It argues, in short, that people do their best under prevailing circumstances. Duncan Snidal defines rational choice as 'a methodological approach that explains both individual and collective (social) outcomes in terms of individual goal-seeking under constraints' (Snidal 2002: 74). Analysing the application of the rational choice theory to EU politics and European integration, Mark Pollack concludes that both the rational choice theory and the constructivist theory, that is the theory of the international system as a social construction, 'have contributed to the development of a sophisticated research program on EU socialization, as well as pioneering (alongside historical institutionalists) a revitalized discussion on the endogenous sources of change in the EU' (Pollack 2007: 50). Pollack argues that rational choice theory is particularly appropriate for 'theorizing endogenous preference formation, particularly insofar as national preferences are derived from the aggregation of individual preferences through domestic political institutions' (Pollack 2007: 47).

The minority Europeanization literature analyses predominantly the impact of EU integration on the change of domestic politics towards national minorities (Hughes and Sasse 2003, Hughes, Sasse and Gordon 2005, Toggenburg 2004). Compared to the first-generation minority rights conditionality, applied towards Central and East European accession countries, the second-generation conditionality extends the minority criterion to the necessity to cope effectively and efficiently with the manifold legacies of the armed conflicts: investigation and prosecution of war crimes, assurance of refugee return and repossession of property. Moreover, the second-generation conditionality appraises the reconciliation attempts in post-conflict societies (Toggenburg 2008). The perspective of EU membership is a powerful reform incentive for the Western Balkan countries. Improvement in refugee return and a credible minority policy, with measures for integration of returnees belonging to national minorities are all important parts of pre-accession conditionality, and as such should have enhanced Croatian prospects for EU integration. Certainly, domestic politics in the pre-accession period, particularly in the negotiations period, became increasingly subjected to European policy expectation towards refugee return. Surprisingly, such an element of 'soft' *acquis*, lacks any normative basis, since it is not, under any circumstances, part of the EU legal corpus. The introduction of refugee return as a pre-accession criterion confirms the academic argument of the phenomenon of Europeanization being conceptualized fairly vaguely (Knill 2001, Radaelli 2003, 2004). Therefore, an accomplished change and reform results in this particular area should be operationalized at the policy level (Börzel and Risse 2000) rather than at the institutional level (Knill 2001). This chapter will therefore explain actor preferences vis-à-vis effective refugee return as an integral part of a minority protection criteria in the most recent and ongoing EU enlargement waves.

The (In)Effectiveness of a Soft Refugee Return Conditionality in the Case of Croatia

Croatia is the only country of the region that has completed the membership negotiations and signed the Accession Treaty. In spite of the fact that Macedonia was granted candidate status back in 2005, that country has not yet started negotiations since a dispute with neighbouring Greece over its official name has held up its EU membership application. In the meantime, Montenegro and Serbia have won candidate status, in 2010 and 2012 respectively.

The EU recognized Croatia as a sovereign state in January 1992 but considered that the political regime of that time was not adequately addressing issues such as democratization, respect of human rights, including minority rights, and the rule of law, and thus had not strengthened relations with Croatia throughout the second part of 1990s. The commitment of the Croatian governments and the presidents who served after 2000 to the democratic values and principles promoted in the tailor-made political conditionality for Croatia and other Western Balkan countries (that include the respect of human rights, including minority rights, prosecution of war crimes, progress in refugee return, and improved relations with neighbouring countries) has in the Croatian case resulted in steady and substantial development of relations with the European Union, particularly in comparison to other Western Balkan states.

A clear cross-party consensus on EU membership as a supreme Croatian foreign policy goal has been maintained since 2000. In this way, the desired membership of the European Union encouraged Croatian political elites to foster friendly relations with neighbouring countries, and to remain committed to the return of refugees from neighbouring countries to Croatia, resulting in adjustment of the domestic normative framework and policy solutions to meet those political goals. A Resolution on the Accession of the Republic of Croatia to the European Union adopted in December 2002 by all political parties in the Croatian Parliament defined Croatia's accession to the EU as a strategic national goal and recommended the government to submit an application for EU membership (Croatian Parliament 2002).

Informally, Croatia's EU accession process started in 2000 at the Zagreb Summit and was formalized on 29 October 2001 when Croatia signed a Stabilisation and Association Agreement (SAA) with the EU. The SAA does not include a direct reference to the necessity to assure minority rights in the accession process (Engl and Harzl 2009: 322), but minority protection became an issue of concern when the European Commission prepared the Opinion on Croatia's EU membership application. The government submitted the application for EU membership on 21 February 2003. Later that year, at the European Council summit held in Thessaloniki in June 2003, all of the Western Balkan countries were given the prospect of EU membership as long as they fulfilled the necessary conditions. In its Opinion on Croatia's EU membership application from 20 April 2004, the European Commission recommended the opening of EU accession negotiations. In its Opinion, the Commission considered that Croatia needed to take further measures to ensure that the rights of minorities, in particular of the Serb minority, were fully respected. Croatia was furthermore advised to speed up the implementation of minority-related legislation and to accelerate efforts to facilitate the return of Serb refugees from Serbia and Bosnia and Herzegovina (European Commission 2004: 37). The Commission recommended in the Opinion of 14 April 2003, that negotiations for accession to the EU should be opened in March 2005, provided that the country had demonstrated full cooperation with the ICTY (European Commission 2004).

The Opinion was accompanied by the European Partnership for Croatia, which listed short- and medium-term priorities for Croatia's preparations for further integration with the European Union (EU Council 2004). The European Partnership reflected the country's stage of preparation at the time it was passed and was tailored to fit its needs. Croatia was expected to respond to the European Partnership by preparing a plan with a timetable and details in terms of how it intended to address the European Partnership's priorities. In other words, the Partnership served as a checklist against which to measure progress. It established measures for the Copenhagen political criteria concerning human rights and protection of minorities. The completion of refugee return and accomplishment of housing reconstruction was among the short- and medium-term priorities. Croatia was requested to create social and economic conditions to improve the climate for returnees and the acceptance of returnees by receiving communities. The Partnership also formulated the necessity for enhanced regional cooperation to accelerate the process of refugee return. As a medium-term priority, it stated that the process of refugee return should be completed through proper and timely implementation of relevant legislation. In addition, the economic and social reintegration of returnees should be ensured through regional development programmes in previously war-affected areas.

Nevertheless, due to insufficient cooperation with the ICTY, the accession negotiations were suspended until 3 October 2005. Negotiations were opened after the assertion of Croatia's general fulfilment of the political criteria, established by the EU Summit in Copenhagen, in 1993. The Commission stated clearly that the negotiations would be based on Croatia's own merits and the pace would depend on Croatia's progress in meeting the requirements for membership, thus not linking the Croatian negotiating procedure to any other applicant country's. The Chapter on Judiciary and Human Rights, in which minority protection and refugee return were to be assessed, was opened in September 2006 when the first bilateral screening meeting took place. Within the negotiation process, the Commission requested the Croatian authorities to provide data on the actual number of registered returnees and the concrete measures undertaken in order to implement the respective human rights of the refugees. Consequently, the authorities were asked to develop action plans in which progress in implementation of certain elements of minority protection had been assessed. However, refugee return was not one of the issues the Commission was particularly keen on scrutinizing.

Throughout the Croatian accession process, the European Commission has assessed the implementation of conditionality in its yearly progress reports. These reports evaluate the country's achievements in the implementation of the EU *acquis communautaire*, and thus its preparations for EU accession, over the past 12 months, providing a concise track record of the country's progress. The reports include the analysis of the situation in the country with respect to the Copenhagen political criteria for accession: the existence of stable democratic institutions guaranteeing a democratic state order; the rule of law; human rights; and, last but not least, the country must prove its respect for, and protection of, its minorities. In addition, the EC scrutinized the progress made in respect to the return of refugees and repossession of their pre-war properties, both in the section of the progress report dealing with political conditionality and in the one dealing with regional issues and international obligations. Throughout these years the Croatian authorities were encouraged to assure full implementation of minority legislation and to maintain commitment to implementing all obligations related to the return of refugees and internally displaced persons.

The 2005 Progress Report acknowledged that the position of the Serb minority in Croatia had improved since the Opinion, partially because of progress in terms of repossession and reconstruction of Serb-owned housing (European Commission 2005: 22). The Report recorded a substantial decrease in the number of refugees from Croatia, compared to previous official figures in Serbia and Montenegro and Bosnia and Herzegovina at the end of 2004 and the beginning of 2005. According to the results of reregistration, the number of refugees from Croatia in Serbia and Montenegro as of July 2005 was 101,875, while it was 7,566 in Bosnia and Herzegovina, down from a total of approximately 208,000 in both countries in May 2004. In the same year, the Report revealed that the total number of Croatian Serbs registered as returnees to Croatia was 96,892 or just over one-quarter of those who left the country during the war. Approximately 85,002 of these returned from Serbia and Montenegro and 11,890 from Bosnia and Herzegovina. It would appear, based on the latest figures, that a large number of Serb refugees have opted for local integration in the country they reside in. It should also be noted that many of the 96,892 officially registered returnees to Croatia have subsequently left the country. Estimates point to a sustainability rate of return of around 60–70 per cent. Apart from referring to several ethnically motivated incidents that occurred in the course of the previous year, the Report underlined that there were 'still real obstacles to the sustainable return of Serb refugees, such as enduring, hostility in certain localities, and remaining housing concerns, including those involving former tenancy rights holders' (European Commission 2005).

The 2006 Progress Report states that around 5,000 had returned over the past year. The total number of Croatian Serbs registered as returnees to Croatia in that year was 126,573, or about one-third of those who left the country during the war. Estimates point to a sustainability rate of return of around 60–70 per cent. The Report acknowledged that there were also still real obstacles to the sustainable return of Serb refugees, such as enduring hostility in certain localities, and remaining housing concerns, mainly those involving former tenancy rights holders (European Commission 2006: 13). The European Commission warned that political will is crucial factor 'if the housing care programmes are not to remain largely an empty promise' (European Commission 2006: 17).

In 2007, the Commission reported mixed developments with regard to the Serb minority and acknowledged that the mood in the country continued to move forward, albeit slowly (European Commission 2007: 13). Limited progress on the various outstanding issues regarding refugees was noted, emphasizing that around 3,500 refugees returned to Croatia in the course of the year. The total number of Croatian Serbs registered as returnees to Croatia increased to 130,000, although the estimated level of actual return could be less than 60 per cent of this figure. The Commission warned that no substantive decision had been taken to settle the issue of convalidation of certain documents and rights (including pension rights, of those residing in the parts of Croatia not under Croatian government control during the 1990s), employment opportunities and enduring hostility in certain return areas (European Commission 2007: 14).

The 2008 Progress Report considers the appointment of an ethnic Serb to one of the Deputy Prime Minister posts as a positive development (European Commission 2008: 13). It acknowledges some progress on the various outstanding issues regarding refugees, reporting on more than 2,100 refugees and 1,100 displaced persons who returned to their place of origin in Croatia in 2007. The total number of Croatian Serbs registered as returnees to Croatia increased to almost 143,000 although the estimated level of actual return could be less than 60 per cent of this figure. The report, however, underlined that a number of obstacles to sustainable return of Serb refugees remained, principal among them being housing, particularly for former tenancy rights holders (European Commission 2008: 14).

The 2009 Progress Report mentioned 'some progress on the various outstanding issues regarding refugees', such as in providing housing solutions to returning refugees, in reconnecting public infrastructure in certain return villages and in mine clearance operations. The report nevertheless stated that there were still a number of obstacles to sustainable return of Serb refugees, the main one being housing, particularly for former tenancy rights holders. The Commission criticized that fact that implementation of the Croatian government's housing care programmes was lagging behind the planned timetable. In that year, around 1,345 refugees and 380 displaced persons returned to their place of origin in Croatia. The total number of Serb minority returns to Croatia at that point stood at 108,466, in addition to 23,231 internally displaced Serbs who had returned to their homes (European Commission 2009: 15).

Similarly to previous reporting periods, the 2010 Progress Report noted 'some progress on the various outstanding issues regarding refugees'. According to the Report, in the past year the Croatian authorities had now registered over 132,600 returnees in total belonging to the Serb minority, which corresponds to approximately half of those who fled the country up until 1995. In addition, the Report acknowledged that the government had met its self-imposed 2008 benchmark of providing 1,400 housing units to former occupancy and tenancy rights holders (European Commission 2010: 14).

Good progress on refugee-return issues was reported in the 2011 Progress Report since 'refugees continued to return to Croatia, with the Croatian authorities having registered over 132,872 returnees in total belonging to the Serb minority, which corresponds to approximately half of those who left the country. It is estimated that 54 per cent of returnees have remained in Croatia. Progress was also noted with regard to housing issues' but acceleration of 'efforts to create the economic and social conditions necessary for sustainable returns' was requested from the government. Overall assessment was that the position of minorities in Croatia had continued to improve, and that good progress was being made on the outstanding refugee-return issues, but additional efforts to provide housing and ensure the sustainability of returns was requested (European Commission 2011: 13).

Progress Reports in the period 2005–2008 had continuously drawn attention to the discrimination and underrepresentation of minorities, particularly Serbs, in access to employment, especially in the war-affected areas. The Accession Treaty was signed in December 2011, with the continuation of the strengthening of minority protection, including through effective implementation of the Constitutional Law on the Rights of National Minorities, listed among the specific commitments undertaken by Croatia in the accession negotiations.

Peculiar Trends in Refugee Return in Croatia

An armed conflict with the local Serb rebels and the Yugoslav People's Army commenced in 1991, resulting in occupation of one-third of Croatia's territory and displacement of the population. Between 1991 and 1997, around 950,000 Croatian citizens were displaced, both within the borders of the Republic of Croatia and outside them. Around 550,000 displaced persons were mainly of Croatian ethnicity, while the remaining 400,000 were minority Serbs, 330,000 of whom were displaced in Serbia and Montenegro, 40,000 in Bosnia and Herzegovina and 32,000 in the Croatian Danube region (Mikić 2008: 2). Nevertheless, a systematic analysis of the refugee-return drawbacks has not yet been done in the Croatian academic literature. This issue has initially been documented and researched vis-à-vis the refugees and internally displaced persons of Croatian ethnic origin (Magdalenić 1994, Mesić 1992, 1994, 1995, 1996, Rogić 1995, Varšek 1999, Vlada Republike Hrvatske 1995), whereas the return of Serb refugees has emerged relatively recently as a research topic (Babić 1998, 1999, 2002, 2004, Mesić and Bagić 2007, 2008, 2012, Radić 2004, Živić 2003).

Whereas the 1991 population census showed that almost 22 per cent of the population was not of Croatian ethnic origin, the 2001 census reported that around 90 per cent of the inhabitants were ethnic Croatians (Croatian Bureau of Statistics 1992: 64). In other words (and with due diligence regarding interpretation of statistical data), while the percentage of ethnic non-Croats had decreased by over 50 per cent within ten years, the number of ethnic Croats had in the same time increased by more than 12 per cent, showing ethnic homogenization during the era of armed conflict and ethnic cleansing. The results of the census further reported that 7.47 per cent of the population belonged to one of Croatia's 22 national minorities, which was half of the total number of the minority population in 1991 (Croatian Bureau of Statistics 2001). The largest autochthonous minority group, the Serbs, have traditionally constituted a majority in the Krajina region surrounding the south-western city of Knin (Zadar and Šibenik-Knin County) and made up a significant proportion of the population in the eastern part of Croatia (the Osijek-Baranja County and the Vukovar-Srijem County). Today, the largest number of Serbs live in the counties of Vukovar-Srijem, Osijek-Baranja, Sisak-Moslavina and Karlovac. A certain number of the Serb minority also dwell in Primorje-Gorski Kotar County, Lika-Senj County and in the City of Zagreb. The number of the Serb minority in Croatia dropped, in the first instance, due to the exodus of the Serb population at the beginning of the 1990s, when small numbers of Serbs left Croatia straight after the country's declaration of independence in 1991. On the other hand, the biggest exodus took place in 1995 following two military actions undertaken by the Croat government to reintegrate the occupied territories in the spring and summer of 1995 (Žunec 1999). The majority of them became refugees in the neighbouring (then) Federal Republic of Yugoslavia (FRY) and Bosnia-Herzegovina. Those military operations resulted in control of all occupied territories with the exception of the Croatian Danube Region (Eastern Slavonia), which remained under UN protection until January 1998. The Erdut Peace Agreement of November 1995 provided for the peaceful reintegration of Eastern Slavonia into Croatia, which was completed in January 1998.

At first, the process of Serb refugees' return and integration has predominantly been slow and inadequate, as the changed ethnic picture was pleasing the political elites of that time. Authoritarianism and resilience to Western liberal values at the highest political level that were, *inter alia*, expressed through resistance to refugee return, contributed to the international isolation of Croatia in the late 1990s (Boduszyński 2010: 105). The exodus of the Serb refugees might be perceived as 'an outcome of nation-state formation' (Koska 2008: 195). According to Aristide Zolberg, 'these processes were not the outcomes of rational economic goals or calculations of the nation-states. On the contrary, emerging states were readily prepared to sacrifice their economic interest for mere political and ideological goals, driving out portions of their population even though it might be against their economic interests' (Zolberg cited in Koska 2008: 195). A vast number of abandoned private properties and a significant decrease in the population in the areas of return, even now, 17 years after the massive exodus of the Serb population took place, speak as an example of the economic failure of Croatian political elites to stop depopulation of part of the country's territory as much as they do of a human rights violation of the refugees.

Surprisingly, the peak of the refugee return happened between 1997 and 2000, when the security situation was still troublesome (Mesić and Bagić 2012: 27). Namely, in 1998 the Return Program was introduced and only after that policy decision of the Croatian government did 'significant, but still very low numbers of Serbian refugees [begin] to return' (Koska 2008: 192). Returnees who returned in this first return wave often originated from predominantly rural areas that were confronted with the problem of land mines and other explosive remnants of war (ERW),[1] and they faced discrimination in access to markets for agricultural

1 According to the 2010 Report of the Council of Europe's Commissioner for Human Rights, extensive areas in Croatia, including agricultural land and forests and more than

produce, as well as very poor road access, electricity and water supplies (Allen, Li Rosi and Skeie 2010: 22). Returnees were also faced by the hostility of local Croatian communities (Koska, 2008: 192).

The introduction of the legal framework on property repossession for refugees and, even more so, its implementation, had been a slow and very costly process. It appears that progress in this field has very likely been terminated. Mesić and Bagić, for example, state that 'the restitution of ownership of homes, land and other real estate to their legal owners (minority refugees) was for the most part accomplished by 2005' (Mesić and Bagić 2012: 30). For example, in 2009 only 24 such cases were pending before Croatian courts (Allen, Li Rosi and Skeie 2010: 23), and out of 19,280 private homes registered as occupied, 19,257 had been returned to their lawful owners (Allen, Li Rosi and Skeie 2010). In April 2010, 22 cases were still pending before domestic courts (Council of Europe 2010). The same cannot be said in terms of *reconstruction* of refugees' property. Private property has not been returned in 13 'unsolicited investment cases' still pending before the courts (US Department of State 2012: 8). These numbers serve as an indication of the ethnic favouring treatment and confirm discrimination towards the Serb returnees in their claims for housing reparation and repatriation. That is, out of approximately 150,000 reconstructed family houses, 35 per cent were for Serb returnees (UNHCR 2010: 2, Mesić and Bagić, 2012: 30). Surprisingly, the reconstruction of houses damaged or destroyed during the war was assessed as a relative success story (Human Rights Watch 2006: 30).

Apart from this, refugee return was hindered by a discriminatory legal framework and by the nonexistent or inadequate political will of both central and local policy makers (Blitz 2003). Serbs who returned in this first return wave often faced long, complex and discriminatory bureaucratic procedures. For example, while ethnic Croats who were refugees to Croatia from neighbouring countries, and who had not previously been residents in Croatia, were able to obtain citizenship quickly, returning Serbs encountered difficulties in proving their right of residence and citizenship. Moreover, a large number of ethnic Serb-owned homes were either destroyed in the war or occupied by ethnic Croats. Finally, a category of former residents who had had occupancy rights to socially owned apartments (*stanarsko pravo*) was for a long time prevented in their return attempts. This category of flat occupancy, which existed under the socialist system, allowed them to enjoy a life-long tenancy and even to pass the tenancy right on to their children. The abolishment of tenure and tenancy rights has rendered a majority of refugees

---

100 towns and municipalities, remained contaminated by mines and ERW by June 2010. Between 1999 and 2008, more than 200 mine/ERW casualties were reported, including 75 deaths and 137 injuries. It is worth noting here that apart from the security aspect, this issue has a serious economic dimension, which more often than not remains unnoticed in the discourse about land mines as the topic is dominated by security concerns: e.g. the loss to agriculture is estimated at €44 million a year while the value of wood resources that cannot be used due to land mines is estimated at €178 million (Council of Europe 2010).

with limited and low incomes more vulnerable to sub-standard living conditions, evictions and homelessness (Mikelic and Schoen 2005).

In their study, Mesić and Bagić argue that 'occupancy and tenancy rights certainly do not belong to the field of international refugee law, but they may be considered humanitarian law for reasons of obvious discrimination against Serb refugees at the time of their abolishment' (Mesić and Bagić 2012: 26). Namely, when Serb refugees left their homes during the war, the authorities took up the right to repossess their houses and made them available to refugees, settlers and displaced people, primarily of Croat ethnicity (Council of Europe 2010). Those activities had a legal ground in the 1995 Law on Temporary Take Over and Administration of Specified Property and the Law on Lease of Flats and the Law on Lease of Flats on Liberated Territories that abolished occupancy and tenancy rights of 'socially owned' flats (Council of Europe 2010: 31). Although this legislation that allowed for violation of refugees' private property was repealed in 1998 at the insistence of the international community, the repeal had no practical effect as the termination of specially protected tenancies had already been accomplished (Human Rights Watch 2006: 5). In the meantime, the new residents, mostly refugees of Croatian origin, were given the right to buy those properties on favourable terms, effectively blocking the possibility of the former residents returning to their original homes. The Croatian authorities started addressing this issue and established a compensation programme for former occupancy and tenancy rights in June 2003, when the Housing Care Programme for Former Occupancy and Tenancy Right Holders was adopted (Human Rights Watch 2006).

Based on 2002 Amendments to the Law on the Areas of Special State Concern (ASSC) and a Government Decision of June 2003, publicly financed housing care became available both within and outside the ASSC for those former tenancy right holders who expressed their intention to return (Mikelic and Schoen 2005: 83). Whereas no deadline was set for applications inside the ASSC, former tenancy right holders outside the ASSC were given a time limitation in applying for the publicly financed housing care programme. The deadline for submission of their requests, originally set for the end of 2004, was extended several times on the request of the Serb political representatives as well as the suggestions of the international community. Subsequently, out of 14,006 family applications for housing care from former occupancy and tenancy right holders submitted to the Croatian authorities, 7,853 have so far been positively resolved (Mikelic and Schoen 2005: 26). Mostly under international community pressure, the Croatian government accepted that they needed to find a solution for the former occupancy and tenancy right holders, allowing them to legally purchase their formerly used apartments under favourable conditions and continuously extending deadlines for filing new housing care applications (Mikelic and Schoen 2005).

In total, 132,872 people had been registered by the Croatian authorities as minority returnees (European Commission 2012: 13). Compared to the original caseload of 297,000 refugees from Croatia registered at a certain point of time only in neighbouring Serbia (Allen, Li Rosi and Skeie 2010: 18), the number of sustained returns seems relatively small. Moreover, it is estimated that 54 per cent of returnees have remained in Croatia. Acknowledging the right of every refugee to voluntarily decide their preferred place of residence, and on the grounds of the permanency of the stay of minority returnees in Croatia, Mesić and Bagić distinguish between two broad categories of returnees: real ones and formal ones. Their empirical research established that every third registered returnee lives in Croatia, more or less on a permanent basis (real or permanent returnees). Almost 40 per cent of registered returnees do not live continuously in the territory of Croatia (formal returnees), of which the majority (70 per cent) live in Serbia. Around 15 per cent of registered returnees passed away before their research was conducted, indicating that the actual number of registered returnees the government has been operating with might be even smaller. Finally, for about 13 per cent of the selected sample of returnees, it was not possible to collect reliable data on the place of their stay. The authors of the study stress that this group particularly makes it difficult to make precise estimates of the proportion of real and formal returnees (Mesić and Bagić 2012: 81–2).

Housing conditions and difficulties related to the restitution of property, and the slow pace of and difficulties in housing reconstruction are just part of the problems returnees have faced. Recent research on sustainable return concluded that subjective issues are just as important an indicator of quality of life as infrastructure problems or economic conditions (Mesić and Bagić 2007, 2012, Koska 2008). The research found that the number of people who are dissatisfied is reducing, and the proportion of those who are very satisfied is increasing with the increase in total household income (Mesić and Bagić 2012: 97). Comparing refugees in general to life's losers, Mesić and Bagić realistically assessed that 'the great majority of refugees and returnees experience losses, whether economic or social, in comparison to their earlier circumstances, although we should bear in mind that some (formal) refugees experienced political and material gain as a direct result of ethnic conflict and its social consequences' (Mesić and Bagić 2012: 179).

The household income of returnees remains significantly below that of the population in general (Allen, Li Rosi and Skeie 2010). The majority of unfavourable employment opportunities arise from weak economic development in the areas of return belonging to so-called Areas of Special State Concern that are home to 15.3 per cent of the total population in Croatia (Lovrinčević, Mikulić and Budak 2004). Mesić and Bagić explain that now that security is no longer a concern, infrastructure deficiencies closely connected to bad socioeconomic conditions are discouraging (particularly younger) people from returning. In their words:

> [l]iving conditions in small places that are isolated from the transportation perspective are certainly not attractive to younger generations of refugees, but not even to the elderly who have become used to a different, higher living standard during their time as refugees. It should therefore be no surprise that some, particularly younger returnees give up on returning to such conditions or

> they do make the decision to not to return at all. Their decisions do not depend very much on political circumstances, inter-ethnic relations or other topics that are frequently discussed in the context of refugee return and stay. Many of them would have probably left their places of birth even if there had not been for the war, in pursuit of better living conditions and better opportunities. Creating better living conditions in such areas and regions is the issue of general economic and urban development of underdeveloped parts of Croatia, and it cannot be resolved with special measures or policies focused solely on issues of return and returnees. (Mesić and Bagić 2012: 99–100)

Therefore, a coherent and effective regional policy is of vital importance, not only for achieving social cohesion across Croatian regions, but also for the preservation of the minority population in return areas (Lika, Banovina and Kordun, and Dalmatia). Public attitude towards the return of Serbs has improved but better employment opportunities would even more concretely help inter-ethnic reconciliation. The government has attempted to address the lack of economic opportunities in the areas of return and economically underdeveloped regions through favourable initiatives for entrepreneurs in the so-called Areas of Special State Concern. Such initiatives related to war-affected areas include tax advantages and other incentives to encourage investment and job creation. These measures do not contain any measures for positive discrimination of minorities or returnees, however. In most areas of return, Serbs are rarely employed in public services and local administration. Discriminatory employment practices towards the Serb minority were a matter of concern of numerous reports on implementation of minority rights in Croatia (Council of Europe 2005, Human Rights Watch 2003, International Crisis Group 2002, Minority Rights Group International 2003).

It has been mentioned above that occupancy and tenancy right of refugees are a legacy of the former socialist system. As such, they are not part of the conventional international refugee law. Nevertheless, the international community has linked restitution or compensation for refugees' occupancy and tenancy rights to post-conflict resolution policies (including the EU conditionality), considering this 'as a (pre)condition for the return of refugees and displaces, especially the minority members, in Bosnia and Herzegovina, Croatia and in Kosovo' (Mesić and Bagić 2012: 31). Responding to the situation of refugees deprived of such tenancy and occupancy rights, the Croatian government has put forward the Housing Care programme. This provides accommodation to people in need of it, including the right to buy at preferential rates in war-affected areas. The beneficiaries of this programme include refugees, internally displaced persons and returnees, as long as they have the intention to settle permanently in Croatia and do not have property rights elsewhere. The Action Plan envisages the formation of joint commissions with the United Nations High Commissioner for Refugees (UNHCR), the EC, the Organization for Security and Cooperation in Europe (OSCE) and the USA, in order to ensure that the Housing Care programme is transparent, credible and speedily implemented. Significantly, the effectiveness of the programme is one of the key benchmarks being monitored for the EU accession process (Allen, Li Rosi and Skeie 2010).

Apart from economic conditions, political and social circumstances also influence the success of the return process. At the political level, a change in the government's treatment of national minorities took place after the change of the country's executive power in 2000 (European Commission 2004: 8, Petričušić 2004), which also resulted in improvement of relations between the EU and Croatia. The Commission acknowledged the determination of the new political leadership 'to entrench democratic values and principles and made rapid and significant progress on the main political outstanding questions, notably the respect of human rights, including minority rights, democratization of the media, full compliance with the Dayton/Erdut Agreements, cooperation with ICTY and improved relations with neighbouring countries' (European Commission 2004: 8).

The Croatian legal framework for the protection of national minorities is in line with the highest European standards, since Croatia is a party to the Council of Europe's Framework Convention for the Protection of National Minorities and the European Charter for Regional or Minority Languages (Petričušić 2004, Tatalović, 2005). The standards set in those international legal sources have to a great extent been enshrined in the 2002 Constitutional Law on the Rights of National Minorities (Petričušić 2003). Two former coalitional governments included a Serb party, the Independent Democratic Serbian Party, as a minor coalition partner. This could have provided an even more significant incentive to address the situation of the Serb minority, especially refugee returnees of Serb origin. However, despite the progress achieved in implementation of the European minority rights instruments' standards and Croatia's enduring commitment to protection of national minorities, the Committee of Ministers of the Council of Europe pointed in 2011 to a number of remaining issues of concern and to reported cases of discrimination of persons belonging to the Serbian minority, particularly in the fields of employment, housing, recognition of property and other acquired rights. Further, the Council of Europe's Committee of Ministers has directed attention to the problems in regard to the reconstruction of housing units damaged during the war, the sustainability of the return of persons belonging to national minorities, ethnically motivated incidents against persons belonging to national minorities and difficulties in obtaining Croatian citizenship for some returnees, and has recommended further measures to be taken by the authorities towards improvement of the state of minority rights (Council of Europe 2011).

Role of International Human Rights Watchdogs in Endorsing Refugee Return

The Dayton/Paris Peace Agreement of December 1995, which ended the war in Bosnia and Herzegovina, confirmed the right of all refugees to return to their homes. Croatia is, together with Federal Republic Yugoslavia of that time and Bosnia and Herzegovina, a signatory party to this agreement. The OSCE and the UNHCR worked particularly on realization of the right of refugees to return to their pre-war homes and reclaim their property throughout the region. In Bosnia and Herzegovina, the Office of the High Representative, together with a number of international agencies present in that country, 'closely reviewed the revocation of the discriminatory war legislation and actively supported the adoption and implementation of adequate housing laws, which encouraged refugees and internally displaced persons to return to their pre-war homes' (Mikelic and Schoen 2005: 1).

The international community has played an active role in finding durable solutions for refugees and displaced persons in the region and each of its countries. For example, the OSCE Mission to Croatia and the UNHCR office have overseen progress in refugee return, explaining it as an issue belonging to international obligations that the Croatian state (under several governments) had voluntarily subscribed to, including the commitment to implement the obligations effectively and efficiently. Apart from the two aforementioned international agents, the European Union, as well, is closely engaged in promoting just, comprehensive and durable solutions to the problems of refugees and IDPs in the Western Balkans.[2]

Whereas, at the beginning of the post-conflict rehabilitation processes following the armed conflicts of the 1990s, the UN, the OSCE and the Council of Europe were clearly the most active (and visible) actors in the region, in terms of financial and human resources as well as at the political level through dialogue and cooperation with the countries of the region in multilateral forums, over the past 15 years the EU has become the most relevant international actor for reform, as the countries of the region increasingly – and increasingly determinatively – step on the path towards EU integration. (In the early years the EU was more active in funding reconstruction and was much less visible than other actors, but it has since become the main reference point for the countries of the region in all relevant reform areas, including in human rights implementation and promotion.) This development is linked to the fact that the UN, and even more so the OSCE, as purely political organizations with no legally binding standards, had weak or ineffective leverage tools and thus were lacking the 'sticks and carrots' necessary to persuade inactive, or (for whatever reason) outright resisting, states to fulfil their international obligations, especially in the area of human rights. The EU, no doubt, possessed such means of leverage to a much greater extent, both in political and economic terms: the prospect of EU membership was and remains (and probably will remain in the future, regardless of the institutional and financial crisis currently facing the EU) the main foreign policy goal of all countries in the

2 See, for example, *EU Statement on the Ministerial Review Conference on Solving the Refugee Situation in the Western Balkans*, 10 November 2011. Available at: http://eeas.europa.eu/delegations/vienna/documents/eu_osce/permanent_council/2011/20111110_877_conference_on_solving_the_refugee_situation_en.pdf.

region, and thus constitutes a strong incentive and excellent catalyst for reform and cooperation. It is also just as much a strong factor of persuasion whenever political will for the implementation of international obligations, including in the human rights area, is lacking, insufficient, slow or inadequate.

Cooperation with international bodies such as the ICTY and Interpol in the hunt for indicted war-crimes fugitives is one thematic case in point, where the EU used its tool of prospective membership (and promotion to various stages in the approximation process, such as candidacy status) in order to 'persuade' political actors to give up previous – perceived or actual – shortcomings (or outright reluctance) in cooperating with said bodies and do more to find the indictees in hiding. The political prospect of EU membership is paralleled with the potential economic benefits, as the EU integration process provides eligibility for and access to a variety of sources of financial assistance and funding, as well as increasing inclusion in the economic cooperation framework of the Union and access to a large market. The Council of Europe Commissioner for Human Rights has linked the implementation of human rights of refugees and displaced persons in the region to the process of ethnic reconciliation in the region and on the development of national systems for the protection of human and minority rights (Council of Europe 2012).

An additional encouragement for the regional dimension in refugee return came with the internationally driven process which recognized the primacy of the individual's right to choose where to live, rather than emphasizing return over other solutions. At the initiative of the international community (the OSCE, the EU and the UNHCR) the governments of Croatia, Bosnia and Herzegovina and the former State Union of Serbia and Montenegro were brought together in October 2004 with the aim of finding an agreement on the remaining open refugee issues. Realizing that refugee return is a policy with a scope broader than a domestic one, it was hoped that such a regional process would add impetus to removing the obstacles to return, eventually creating the conditions for those displaced to make a free choice on whether to return or locally integrate in the country of refuge.

A trilateral declaration was eventually signed by the ministers responsible for refugee matters in the three countries on 31 January 2005 in Sarajevo. The governments committed to the resolution of all practical, political and legal issues affecting refugees by the end of 2006, through meeting concrete benchmarks to assure the non-discriminatory access to rights such as property, pensions, health care, employment and the facilitation of local integration for those wishing to remain in their countries of refuge; to develop national plans (road maps) specifying tasks and benchmarks at a national level, as well as a regional matrix listing tasks to be tackled jointly. An intergovernmental task force was appointed that will meet at least four times a year to oversee the drafting and mutual agreement of the road maps and the regional matrix as well as the implementation of the process. Each of the countries concerned was given a list of tasks considered important for the successful implementation of the Sarajevo Declaration.

In spite of quite regular task force meetings and the development of national road maps (which eventually resulted in the joint matrix), significant measurable progress in refugee return has not been achieved. However, four countries of the region (when Serbia and Montenegro disintegrated in 2006 each country became a part of the Sarajevo Process) have made progress on a number of outstanding issues such as data exchange, civil documentation, public information, pensions and trust fund mechanisms (European Commission 2011: 14). The European Commission itself acknowledged that progress on regional cooperation and implementation of the Sarajevo Declaration was 'slower than expected' (European Commission 2006: 13) and that 'no decisive progress on implementation of the Sarajevo Declaration … has been achieved' (European Commission 2007: 15). According to the Commission, the lack of political consensus between the signatories blocked the process for a long time (European Commission 2007).

The Refugee High Commissioner's 2008 initiative came at a time when the momentum behind this process had started to decline. The UNHCR initiative reinvigorated it, leading to the ministerial meeting in Belgrade in March 2010, at which the foreign ministers of all four countries made a new commitment to the search for solutions (Allen, Li Rosi and Skeie 2010: 1), agreeing to work together to clarify refugee statistics (European Commission 2010: 16). In spite of this renewed commitment, a number of difficult issues remained open, such as occupancy and tenancy rights. In November 2011, the foreign affairs ministers of Bosnia and Herzegovina, Croatia, Montenegro and Serbia signed a Joint Declaration, including a Framework Programme to help resolve the protracted displacement situation in the Western Balkans. A five-year plan agreed there envisages the closing down of all refugee centres and providing housing for all displaced people. On 24 April 2012 a conference on housing solutions for refugees and displaced persons was organized in Sarajevo with the aim of underlining the international community's firm support to close the refugee chapter across the region.

Upon signing the Sarajevo Declaration, the Croatian authorities were instructed by the international community with a list of tasks to be included in the Croatian Road Map. Soon after (in April 2005), the government established a special working group, responsible for the coordination of the drafting and implementation of the Croatian Road Map. A Croatian road map was eventually developed, comprising a series of tasks to resolve all refugee issues, initially planned for the end of 2006. This proved over-ambitious. However, the government was kept vigorous in its implementation of the tasks set in the Road Map. For example, the closure of the OSCE Mission in Croatia was conditioned on completion of the tasks set in the Road Map.

Conclusion

At first, security and institutional obstacles prevented the return and reintegration of Serb refugees in Croatia. Although security concerns ceased to exist a long time ago, refugees of minority Serb origin have not returned in significant numbers. The Croatian authorities registered over 132,872 returnees in total belonging to the Serb minority. It is estimated that 54 per cent of those who registered their return have remained in the country, which results in merely 65,000 real returnees. Interestingly, and paradoxically, the peak of return was as early as the late 1990s, when conditions for returnees were much worse than today. The poor state of the economy in the return areas, resulting in a lack of employment opportunities, constitutes the main obstacle to return today.

Introduction of the refugee-return conditionality cannot be ascribed to a definition of Europeanization as a process in which the EU facilitated horizontal policy transfer in the (potential) candidate countries, since this criterion presents a case of tailor-made conditionality, that, firstly, responded to gross human rights violations that were occurring throughout the 1990s in the Western Balkan region, and secondly, lacked normative resonance in the EU legal corpus. The political implications of failed and inadequate return from the second part of the 1990s transformed into a new condition to be pursued within the EU accession process that Croatia started in 2003. Attaching it to the pre-accession political criteria, particularly to the criterion of respect for and protection of minorities, as well as the Stabilisation and Association-based criterion of regional cooperation, the EC made refugee return one of the conditionality contents for Croatia and other countries of the region. Although the EU conditionality neither increased the number of actual returnees across the region nor brought the population displacement to an end, it produced some positive impact, primarily by assuring continuous support of the region's governments in the return process.

Realizing that refugee return cannot be properly addressed at the level of nation-states, in 2004, the European Union and other international actors put forward a regional solution with the aim of ending the protracted refugee situation in the Western Balkan region. The Sarajevo Process, named after the city in which a Declaration on Regional Refugee Return was signed, gathered the governments of Bosnia and Herzegovina, Croatia, Montenegro and Serbia under the supervision of the international community and hoped to find durable and just solutions for refugees who were displaced as a result of the armed conflicts between 1991 and 1995. Although the Sarajevo Process brought in a joint commitment of the four governments and vigour and financing from the international community, it has so far produced only limited effects and has not significantly encouraged a rise in the number of returnees. Nevertheless, the European Union membership aspiration has certainly kept the four governments of the region committed to refugee return. Being continuously raised as part of political pre-accession criteria in the EU accession process, refugee return became the 'soft' pre-accession political criterion, and the EU accession process continues to serve as a means of rectifying the consequences of population displacements through its conditionality developed for the Western Balkans.

References

Allen, R., Li Rosi, A. and Skeie, M. 2010. Should I Stay or Should I Go? A Review of UNHCR's Response to the Protracted Refugee Situation in Serbia and Croatia. Geneva: United Nations High Commissioner for Refugees. Available at: http://www.unhcr.org/4d08e19a9.pdf [accessed: 10 June 2012].

Babić, D. 1998. Povratak i mogućnost suživota ratnih stradalnika u zapadnom dijelu Brodsko-posavske županije. *Zlatna dolina*, 4(1), 271–85.

Babić, D. 1999. Sukobi i suradnja povratnika i useljenika u poslijeratnom razdoblju: Brodsko-posavska županija. *Migracijske teme*, 15(4), 483–500.

Babić, D. 2002. Oprost i pomirenje kao pretpostavka suživota: Proces koji je počeo ili utopijski izazov? (Slučaj Brodsko-posavske županije). *Revija za sociologiju*, 33(3–4), 197–211.

Babić, D. 2004. Suživot Hrvata i Srba u prijeratnom, ratnom i poslijeratnom razdoblju implicitna kritika interpretacije rata u Hrvatskoj kao etničkog sukoba. *Migracijske i etničke teme*, 20(2–3), 187–208.

Bieber, F. 2007. Less Diversity – More Integration: Interethnic Relations in the Contemporary Balkans. *Southeastern Europe*, 32(1), 23–38.

Blitz, B.K. 2003. Refugee Returns in Croatia: Contradictions and Reform. *Politics*, 23(3), 181–91.

Boduszyński, M.P. 2010. *Regime Change in the Yugoslav Successor States: Divergent Paths Toward a New Europe*. Baltimore, MD: Johns Hopkins University Press.

Börzel, T. and Risse, T. 2000. When Europe Hits Home: Europeanization and Domestic Change. *European Integration online Papers*, 4(15). Available at: http://eiop.or.at/eiop/pdf/2000-015.pdf [accessed: 10 June 2012].

Council of Europe 2005. *European Commission against Racism and Intolerance: Third Report on Croatia (Released on 14 June 2005)*. Available at: http://www.coe.int/t/dghl/monitoring/ecri/Country-by-country/Croatia/Croatia_CBC_en.asp [accessed: 10 June 2012].

Council of Europe 2010. *Report Following CoE HCHR Visit to Croatia from 6 to 9 April 2010*. Available at: https://wcd.coe.int/ViewDoc.jsp?id=1636837 [accessed: 10 June 2012].

Council of Europe 2011. *Resolution CM/ResCMN(2011)12 on the Implementation of the Framework Convention for the Protection of National Minorities by Croatia*. Available at: https://wcd.coe.int/ViewDoc.jsp?id=1812557&Site=CM&BackColorInternet=C3C3C3&BackColorIntranet=EDB021&BackColorLogged=F5D383 [accessed: 10 June 2012].

Council of Europe 2012. *Post-War Justice and Durable Peace in the Former Yugoslavia*. Available at: http://www.coe.int/t/commissioner/source/prems/Prems14712_GBR_1700_PostwarJustice.pdf [accessed: 1 August 2012].

Croatian Bureau of Statistics 1992. *Statistical Almanac of Croatia 1992*.

Croatian Bureau of Statistics 2001. *Results of the 2001 Census of Population.* Available at: http://www.dzs.hr/Hrv/Popis%202001/popis20001.htm [accessed: 10 June 2012].

Croatian Parliament 2002. *Resolution on the Accession of the Republic of Croatia to the European Union.* Zagreb, 18 December 2002. Available at: http://www.sabor.hr/Default.aspx?art=2491 [accessed 10 June 2012].

Damjanović, I. and Tomičević, L. 2004. Pravni položaj izbjeglica. *Pravnik,* 38(79), 55–84.

Engl, A. and Harzl, B. 2009. The Inter-Relationship between International and National Minority Rights-Law in Selected Western Balkan States. *Review of Central and East European Law,* 34(4), 307–35.

EU Council 2004. *Council Decision on the Principles, Priorities and Conditions Contained in the European Partnership with Croatia.* Brussels, 13 September 2004. 2004/648/EC.

European Commission 2004. *Communication from the Commission. Opinion on Croatia's Application for Membership of the European Union.* Brussels, 20 April 2004. COM(2004)257 final.

European Commission 2005. *Croatia 2005 Progress Report.* Brussels, 9 November 2005. SEC(2005)1424.

European Commission 2006. *Croatia 2006 Progress Report.* Brussels, 8 November 2006. SEC(2006)1385.

European Commission 2007. *Croatia 2007 Progress Report.* Brussels, 6 November 2007. SEC(2007)1431.

European Commission 2008. *Croatia 2008 Progress Report.* Brussels, 5 October 2008. SEC(2008)2694.

European Commission 2009. *Croatia 2009 Progress Report.* Brussels, 14 October 2009. SEC(2009)1333.

European Commission 2010. *Croatia 2010 Progress Report,* Brussels, 9 November 2010. SEC(2010)1326.

European Commission 2011. *Croatia 2011 Progress Report.* Brussels, 12 October 2011. SEC(2011)1200.

Hughes, J. and Sasse, G. 2003. Monitoring the Monitors: EU Enlargement Conditionality and Minority Protection in the CEECs. *Journal on Ethnopolitics and Minority Issues in Europe,* 1(1), 1–36.

Hughes, J., Sasse, G. and Gordon, C. 2005. *Europeanization and Regionalization in the EU's Enlargement to Central and Eastern Europe: The Myth of Conditionality.* Basingstoke: Palgrave MacMillan.

Human Rights Watch 2003. *Broken Promises: Impediments to Refugee Return to Croatia.* Available at: http://www.hrw.org/reports/2003/croatia0903/ [accessed: 10 June 2012].

Human Rights Watch 2006. *Croatia: A Decade of Disappointment: Continuing Obstacles to the Reintegration of Serb Returnees.* Available at: http://www.hrw.org/reports/2006/09/04/croatia-decade-disappointment-0 [accessed: 10 June 2012].

International Crisis Group 2002. *A Half-Hearted Welcome: Refugee Return to Croatia. Balkans Report No. 138*. Available at: http://www.crisisgroup.org/en/regions/europe/balkans/croatia/138-a-half-hearted-welcome-refugee-return-to-croatia.aspx [accessed: 10 June 2012].

Knill, C. 2001. *The Europeanization of National Administrations. Patterns of Institutional Change and Persistence*. Cambridge: Cambridge University Press.

Koska, V. 2008. Return and Reintegration of Minority Refugees: The Complexity of the Serbian Returnees Experiences in the Town of Glina. *Politička misao*, 45(5), 191–217.

Lovrinčević, Ž., Mikulić, D. and Budak, J. 2004. Područja posebne državne skrbi u Hrvatskoj: Razlike u regionalnoj razvijenosti i demografsko-obrazovne karakteristike. *Ekonomski pregled*, 55(5–6), 389–441.

Magdalenić, I. 1994. Prognanici i izbjeglice u Republici Hrvatskoj sredinom 1994. *Revija za socijalnu politiku*, 1(3), 319–21.

Mesić, M. 1992. *Osjetljivi i ljuti ljudi: Hrvatske izbjeglice i prognanici*. Zagreb: Ured za prognanike i izbjeglice Vlade Republike Hrvatske.

Mesić, M. 1994. Pogledi na povratak: Hrvatske i bosansko-hercegovačke izbjeglice. *Društvena istraživanja*, 3(6), 629–49.

Mesić, M. 1995. Tipovi izbjeglica: Hrvatska i bosansko-hercegovačka iskustva. *Društvena istraživanja*, 4–5(18–19), 657–75.

Mesić, M. 1996. *Ljudi na čekanju – pogledi na povratak: Hrvatske i bosansko-hercegovačke izbjeglice i raseljenici*. Zagreb: Sociološko društvo Hrvatske.

Mesić, M. and Bagić, D. 2007. *Održivost manjinskog povratka u Hrvatskoj*. Zagreb: UNHCR.

Mesić, M. and Bagić, D. 2008. Društveno-ekonomski uvjeti održivosti povratka: Primjer srpskih povratnika u Hrvatsku. *Revija za socijalnu politiku*, 15(1), 23–38.

Mesić, M. and Bagić, D. 2012. *Minority Return to Croatia: Study of an Open Process*. Zagreb: UNHCR.

Mikelic, V. and Schoen, T. 2005. *Housing and Property Rights in Bosnia and Herzegovina, Croatia and Serbia*. Nairobi: United Nations Human Settlements Programme.

Mikić, Lj. 2008. *Republika Hrvatska: Izbjegli i raseljeni bivši nositelji stanarskih prava. Osvrt na pristup stečenim pravima i stambeno zbrinjavanje manjinskih povratnika u 2008*. Osijek: Koalicija za promociju i zaštitu ljudskih prava.

Minority Rights Group International 2003. *Minorities in Croatia*. Available at: http://www.minorityrights.org/1005/reports/minorities-in-croatia.html [accessed: 10 June 2012].

Petričušić, A. 2003. Croatian Constitutional Law on the Rights of National Minorities. *European Yearbook of Minority Issues*, 2, 607–29.

Petričušić, A. 2004. Wind of Change: Croatian Government's Turn towards Policy of Ethnic Reconciliation. *European Diversity and Autonomy Papers*, 6.

Available at: http://www.eurac.edu/en/research/institutes/imr/activities/Bookseries/edap/Documents/2004_edap06.pdf [accessed: 10 June 2012].

Pollack, M.A. 2007. Rational Choice and EU Politics, in *Handbook of European Union Politics*, edited by K.E. Jørgensen, M.A. Pollack and B. Rosamond. New York, NY: Sage, 31–56.

Radaelli, C.M. 2003. The Europeanization of Public Policy, in *The Politics of Europeanization*, edited by K. Featherstone and C.M. Radaelli. Oxford: Oxford University Press, 27–55.

Radaelli, C.M. 2004. Europeanisation: Solution or Problem. *European Integration online Papers*, 8(16). Available at: http://eiop.or.at/eiop/pdf/2004-016.pdf [accessed: 10 June 2012].

Radić, A.-M. 2004. Socijalna prava: prognanici, povratnici i izbjeglice. *Revija za socijalnu politiku*, 11(1), 129–36.

Rogić, I. 1995. *Progonstvo i povratak: Ppsihosocijalne i razvojne odrednice progonstva i mogućnosti povratka hrvatskih prognanika.* Zagreb: Institut za primijenjena društvena istraživanja.

Snidal, D. 2002. Rational Choice and International Relations, in *Handbook of International Relations*, edited by W. Carlsnaes, T. Risse and B.A. Simmons. New York, NY: Sage, 73–94.

Tatalović, S. 2005. *Nacionalne manjine u Hrvatskoj* Split: Stina.

Toggenburg, G. von 2004. *Minority Protection and the Enlarged European Union: the Way Forward.* Budapest: LGI/OSI.

Toggenburg, G. von 2008. The Protection of Minorities at the EU-Level: A Tightrope Walk between (Ethnic) Diversity and (Territorial) Subsidiarity, in *European Integration and its Effects on Minority Protection in South Eastern Europe*, edited by E. Lantschner, J. Marko and A. Petričušić. Baden-Baden: Nomos, 83–115.

UNHCR 2008. *Estimates of Refugees and Displaced Persons still Seeking Solutions in South-Eastern Europe, 2008*. Available at: http://www.reliefweb.int/rw/fullMaps_Eu.nsf/luFullMap/EBEA77098B0B628C12574900036510A/$File/unhcr_IDP_eur080331.pdf?OpenElement [accessed: 10 June 2012].

UNHCR 2010. *Policy Development and Education Service: Should I Stay or Should I go? A Review of UNHCR's Response to Protracted Refugee Situation in Serbia and Croatia*. Available at: http://www.unhcr.org/4d08e19a9.pdf [accessed: 10 June 2012].

United Nations 1951. *The 1951 Convention relating to the Status of Refugees*. Available at: http://www.unhcr.org/pages/49da0e466.html [accessed: 12 June 2012].

US Department of State 2012. *The Annual Reports on Human Rights Practices: Croatia*. Available at: http://www.state.gov/documents/organization/186551.pdf [accessed: 10 June 2012].

Varšek, I. 1999. Problemi povezani uz povratak izbjeglica i raseljenih osoba. *Polemos: časopis za interdisciplinarna istraživanja rata i mira*, 2(3–4), 171–217.

Vlada Republike Hrvatske 1995. *Vodič za ostvarivanje prava stradalnika iz Domovinskog rata.* Zagreb: Vlada Republike Hrvatske.

Wendt, A. 1999. *Social Theory of International Politics.* New York, NY: Cambridge University Press.

Živić, D. 2003. Prognano stanovništvo iz hrvatskog Podunavlja i problemi njegovog povratka (1991–2001). *Hrvatski geografski glasnik*, 65(1), 63–81.

Žunec, O. 1999. Operacije Bljesak i Oluja, in *Rat u Hrvastkoj i Bosni i Hercegovini 1991–1995*, edited by B. Magaš and I. Žanić. Zagreb: Jesenski i Turk, 93–111.

Chapter 4

The EU in the Republic of Macedonia: An Active Player, a Framework for Integration or Both?

Simonida Kačarska

This chapter examines the role of the European Union in the Republic of Macedonia between 1997 and 2011 by focusing on the political criteria for EU accession. The chapter utilizes the distinction between the EU as an 'active player' in the mediation and conflict resolution in the region, and the EU as a 'framework', equipping countries with models of governance and policy options in the accession process (Noutcheva et al. 2004). Using this distinction, the chapter argues that, although in the common dynamics of accession it is expected that the EU role would progress from the former to the latter, in Macedonia these two roles have been intertwined. In order to demonstrate this argument the chapter examines the issues the EU has included in the political criteria for accession and demonstrates the increasing EU involvement in areas of high politics traditionally limited to the domestic sphere. The analysis deals with three distinct periods of EU engagement in Macedonia: the Regional Approach of 1997, the Stabilisation and Association period between 2001 and 2004 and the post-2005 period in which Macedonia is a candidate country for accession. The three periods are marked by important milestones in Macedonia's EU accession process. The analysis is based on qualitative methodology and relies on process tracing of official documents and secondary information, as well as data obtained through interviews with the elites involved in the EU accession process conducted in 2010 and 2011.

Introduction

Political, also known as democratic, conditionality 'emphasizes respect for and the furtherance of democratic rules, procedures and values' (Pridham 2002a: 956). In practical terms, it includes the Copenhagen political criteria for accession, but also a range of topics concerning the work of the parliament, government, judiciary, anti-corruption policy, protection of human rights and regional cooperation. Research on the political criteria in the Eastern enlargement has concluded that between 1997 and 2004 the Commission changed its understanding of these conditions, but mostly focused on four issues in its reports: the functioning of the legislature, the judiciary, the executive, and anti-corruption measures (Kochenov 2004). This chapter examines the political conditionality stipulated by the EU towards the Republic of Macedonia from the late 1990s until 2011. By studying the political conditions the EU has put forward, it demonstrates the evolution of the EU's approach in the framework of the regional initiatives at the time: from the launching of the Regional Approach in the late 1990s, to the 2001 Stabilisation and Association Process (SAP) and the post-2005 period in which Macedonia has been a candidate country for accession.

In order to unpack the intertwined role of political conditionality, this chapter combines concepts from the Europeanization and conditionality literature with insight from studies on the EU role in conflict settlement and resolution. While the analysis is linked to the study of Europeanization of the candidate countries, it focuses on the application of conditionality, emphasizing the context-specific nature of this mechanism. Because of the idiosyncrasies of the Macedonian case, the chapter also draws upon a specific strand of Europeanization literature concerned with the role of the EU in bringing about conflict settlement and conflict resolution in divided states, using the distinction between the EU as an 'active player' in conflict resolution and as a 'framework for integration' (Noutcheva et al. 2004).

As a country that has been both a front-runner in the EU accession process and a laggard with extensive EU involvement on the ground, Macedonia is an indicative example for studying the involvement of the EU in the Balkans.[1] In regional terms, the country was the first in the region to sign the Stabilisation and Association Agreement (SAA) and the second (after Croatia) to apply and become an official candidate country for EU accession. However, since 2006, the country's accession process has slowed down for various reasons, including the name dispute with Greece. To keep the country on track for European accession, the European Commission has come up with various proposals such as the visa liberalization process or the high-level accession dialogue (European Commission 2012). At the same time, the EU, through its Delegation and High Representative, has had a quite specific role as a guarantor of a functioning peace agreement between the two major groups in the country and has deployed its first ever police mission in the region. As a result, Macedonia is a litmus test for evaluating the operation of EU political conditionality.

The chapter employs qualitative methodology and relies on extensive use of primary documents and secondary literature. In terms of primary information it uses official EU documents that define the formal political conditions as well as national strategic documents prepared for the European integration process. The secondary literature used deals with the mechanisms of Europeanization and conditionality and the specific conditions of the Western Balkans' EU accession. Lastly, for the on-the-ground empirical analysis the chapter uses data from interviews with stakeholders conducted in Brussels and Skopje in late 2010 and

1 Though with unconsolidated statehood, Macedonia has not been considered as a minimalist state in the work of Bieber on political conditionality (Bieber 2011).

early 2011. Qualitative interviewing of the elites involved in the EU accession process has not been common in studies of the Western Balkans, although this method was dominant in the case of the Eastern enlargement (Grabbe 2006, Hughes et al. 2005, Pridham 2005). The chapter recognizes that 'the empirical focus on countries with a general membership perspective (and thus favourable conditions of EU impact) introduces uncertainty as to whether the findings also hold for non-candidate countries' (Schimmelfennig and Scholtz 2008: 189). Nevertheless, as the objective of the chapter is contextual analysis of the role of the EU, the findings are relevant for the study of the Europeanization of the Western Balkan region.

The chapter is organized in two major sections. Europeanization and conditionality as concepts are examined in the next section, which adopts a process-based understanding of conditionality, highlighting the role of domestic actors in the study of this mechanism. Moreover, it complements the analysis of conditionality with the distinction between the role of the EU as an active player and a framework for integration as devised by Noutcheva et al. Building upon these theoretical insights, the second section of the chapter presents the empirical analysis and tracks the evolution of political conditionality in the case of Macedonia. The empirical analysis is divided into three parts related to the changes in the formal relations between the EU and Macedonia. Starting with the launching of the 1997 Regional Approach, the section moves on to the 2001 signing of the SAA and lastly deals with the post-2005 period. By employing a 'before and after' approach between the different specific periods of EU–Macedonia relations, the objective of the chapter is to highlight the evolution and the intertwining role of the EU through its political conditionality (George and Bennett 2005).

Europeanization, Conditionality and the EU as an Active Player/ a Framework for Integration

Europeanization

Research examining the link between EU accession and the candidate countries has traditionally been framed within Europeanization frameworks. For the purposes of this chapter, Europeanization is understood as 'domestic adaptation to European regional integration' (Vink and Graziano 2007). The majority of studies dealing with this phenomenon have focused on the conditions under which Europeanization takes place, as well as its impact on policies, polity and actors in the first instance on the EU Member States (Risse et al. 2001). The Eastern enlargement of the Union triggered a shift towards research adapting and using this concept on the candidate countries for EU accession (Grabbe 2006, Schimmelfennig and Sedelmeier 2005). In turn, the analysis of the candidate countries has slowly been established as a separate sub-field of a broader research agenda (Sedelmeier 2006). While the Europeanization of the Member States is concerned both with the top-down and bottom-up processes, the analysis in the case of candidate countries focuses on the former (Héritier 2005: 207). As a result, Grabbe underscores the need of an 'analytical framework similar to that used to analyse the EU, but one that takes into account the particular characteristics that were critically different for CEE' (Grabbe 2006: 45). Transition and democratic consolidation stand at the forefront of the processes which affect the specific form of Europeanization in the candidate countries, and this in turn is linked to the accession negotiations (Héritier 2005: 203).

Despite the consensus on the potential of using Europeanization as a framework for the analysis of non-member states, at a more general level, the concept has been criticized primarily for its potential for overstretching (Radaelli 2000). Hence, it has been argued that 'Europeanisation in itself is not a theory, but a phenomenon which a wide range of theoretical approaches have sought to explain' (Bulmer 2007: 47). Specifically in relation to the non-member states, Magen and Morlino criticize the tendency of Europeanization literature towards the homogenization of external influence on domestic reform processes (Magen and Morlino 2009: 12). Furthermore, Hughes et al. consider that a 'fundamental problem for the concept is that macro-level and policy-level studies are inconclusive about the causal effects of "Europeanisation" and demonstrate the persistence of deep structural divergences across national and policy contexts' (Hughes et al. 2005: 28). As a result, in reality, the same set of variables may have a different influence on democratization outcomes depending on the kind of outcome being analysed (Magen and Morlino 2009). Due to these outlined critiques, the study of the role of the EU in acceding countries has been framed within the mechanism of conditionality, examined in the following section.

Political Conditionality

Political conditionality is a mechanism that 'entails the linking, by a state or international organization, of perceived benefits to another state (such as aid, trade concessions, cooperation agreements, or international organization membership) to the fulfilment of conditions relating to the protection of human rights and the advancement of democratic principles' (Smith 2001: 37). Whereas the terms democratic and political conditionality are commonly equated, this paper operates with the latter, aiming to underline the importance of political transformation without the unquestioning inclusion of the democratization aspect. In this respect, Anastasakis highlights that 'EU political conditionality can run counter to democratization, at least in the short term when some of the prescriptions prioritize law and order instead of elections and/or civil society development' (Anastasakis 2008: 366). While taking into account these debates, this analysis operates with the term political conditionality as it deals predominantly with the political criteria for EU accession and their impact on the democratic consolidation of the country under examination.

The dominant literature on conditionality is focused on the conditions for success of this mechanism by examining its impact and outcome. Schimmelfennig and Sedelmeier (2005) have developed three models for the examination of the effectiveness of conditionality: the external incentives model, the social learning model and the lesson drawing model, which were applied in two alternative contexts: democratic and *acquis* conditionality. For them, conditionality is 'a bargaining strategy of reinforcement by reward, under which the EU provides external incentives for a target government to comply with its conditions' (Schimmelfennig and Sedelmeier 2004: 662). They conclude that rule transfer from the EU to the Central and Eastern European Countries (CEECs) and the variation in its effectiveness are best explained by the external incentives model and are linked to the high credibility of EU conditionality and the low domestic costs of rule adoption (Schimmelfennig and Sedelmeier 2005). These findings are qualified by the context-dependent influence of the EU as an actor, which is evident in the extension of their model to the Western Balkans (Schimmelfennig and Sedelmeier 2005: 223).

The external incentive model is centred on rule adoption as the outcome of this mechanism and examines conditionality through the logic of an ideal asymmetric relationship between the EU and the candidate countries. However, it does not deal with the changes of conditionality over time that are at the core of the proposed analysis. In response, Hughes, Sasse and Gordon define conditionality 'by the process of its application rather than by an ideal-type assumed power relationship' (Hughes et al. 2005: 3). Moreover, they argue that EU conditionality 'includes not only the formal technical requirements on candidates but also the informal pressures arising from the behaviour and perceptions of actors engaged in the political process' (Hughes et al. 2005: 2). Hughes et al. distinguish 'between formal conditionality, which embodies the publicly stated preconditions … of the "Copenhagen criteria" and the … *acquis*, and informal conditionality, which includes the operational pressures and recommendations applied by actors within the Commission … during their interactions with their CEEC counterparts' (Hughes et al. 2005: 26).

The process-based approaches highlight the potential for increasing politicization of EU political criteria due to their flexible nature and the subjective approach in their evaluation. This is so, because the flexible nature of these criteria reduces the formal conditionality of enlargement, thereby strengthening the informal conditionality, providing the Commission with greater scope for ambiguity in its policy recommendations (Hughes et al. 2004: 527). In her research on the political criteria, Sasse comes to the following conclusion: '[A]n intense and highly visible international involvement in a politicized issue (or one aspect of them) can produce an overlay of contradictory outcomes: a legal change can hide deeper political or societal problems which might, in fact, have become ingrained in the context of the EU's involvement' (Sasse 2009: 28). In light of these theoretical arguments, this chapter examines conditionality as a process, highlighting the changes it undergoes over time, as well as the increasing potential for politicization in the domestic context.

EU as an Active Player/a Framework for Integration

In addition to using the mechanism of conditionality in the post-Yugoslav space, the EU has been directly involved in the domestic politics of these countries, primarily through negotiating or guaranteeing peace agreements. The variety of roles has been analytically examined by Noutcheva et al. by making a distinction between the EU as 'an active player' and 'a framework' with respect to post-conflict circumstances. In the first case, the EU is an active player involved in mediation and conflict resolution in the region, which gears the involved parties with short-term strategies and provides them with the external push leading towards conflict settlement (Noutcheva et al. 2004). The traditional instruments for accession as a long-term perspective of integration in the EU represent the latter role of 'EU as a framework', which offers the possibility of participation in decision making for these countries and equips them with models of governance and policy options (Noutcheva et al. 2004). Similarly, Berg and van Meurs distinguish in policy terms between both the active impact of the EU in relation to border policies and conflict management in contrast to its passive dimension, which concerns identities and conditionality (Berg and van Meurs 2002: 71). While the EU role as a framework has mainly long-term implications, its function as an active player is intended to have short-term effects on the conflict (Noutcheva et al. 2004).

In light of these overlapping roles of the EU, the Europeanization of post-conflict societies such as Macedonia has been accompanied by the possibility of a creation of a 'superficial layer of common institutions and policy coordination mechanisms which would exist for the purpose of satisfying EU requirements, but would not enjoy domestic support' (Noutcheva et al. 2004: 26). Similar arguments have been put forward by Chandler, who points that 'EU member state-building in the Western Balkans is a clear example of the dangers of the liberal peace approach to post-conflict situations' (Chandler 2008: 529). Chandler further criticizes the extensive role of the EU in the region, arguing that that 'the externally driven nature of the policy process means that political elites seek to lobby external EU actors rather than engage in domestic constituency-building' (Chandler 2008). On the other side of the spectrum, O'Brennan considers that the EU is following its normative power approach to enlargement in the region (O'Brennan 2008: 508). As a normative power, the EU, according to Manners, should act to extend its norms into the international system (2002: 252). Hence, the extensive and multifaceted role of the EU in the region both as an actor and a framework has further raised concerns over the impact on the democratic consolidation of the countries studied.

On the whole, the examination of the EU in Macedonia necessitates a distinction between the two dimensions of EU involvement as an active player and as a framework within the context of the conditionality mechanism. While the first dimension of the EU involvement is directed towards short-term conflict resolution, as the European accession process advances, a shift of role of the EU toward a framework can be expected. Nevertheless, with increasing involvement of the EU in politicized issues (as political criteria usually are) the outcomes can also trigger increased polarization and 'unintended consequences' of conditionality. Using these insights from the literature, the following section reflects empirically on the role of the EU in the Macedonian context by looking at its involvement through the political conditionality mechanism from 1997 until 2011.

The Regional Approach: Nascent Role of the EU

Macedonia has been subject to EU conditionality since the mid-1990s through various instruments and policy frameworks, ranging from the Union's foreign policy to the enlargement portfolio. In the late 1990s, the EU introduced the Regional Approach with the purpose of instigating reforms in the political area, such as return of refugees and inter-ethnic reconciliation (EU Council 1997). Prior to this strategy, every country had established bilateral relations with the Union following their respective recognition by the EC Member States and the establishment of diplomatic relations.[2] The Regional Approach was in fact the first formal EU conditionality for the region (Gordon et al. 2008). Furthermore, these states were to form a new political category under the heading of the 'Western Balkans' (Bartlett and Samardžija 2000). As the launching of the Regional Approach took place before the Amsterdam Treaty came into force, the mechanisms for EU foreign policy coordination were still in a nascent phase of development. In this period, as observed by one think-tank analyst in an interview given to the author, Serbia was still under sanctions and the Kosovo conflict was at its peak, so the role of the EU was shaped through an immediate security threats approach.

With the Regional Approach, 'while refraining from extending the offer of membership, Brussels developed relations with the regional states both on an individual and collective basis. Countries meeting the conditions were to be rewarded with trade concessions, financial assistance, and economic cooperation' (Anastasakis and Bechev 2003). In relation to the tailored dimension, the agreements with the countries were devised taking into account their specific state of development. For example, the 'agreement with Macedonia contains an essential element clause, which is more far-reaching than ordinary clauses. Respect for social rights and maintenance of good neighbour relations are included among the binding conditions' (Fierro 2003: 160). The latter 'collective' dimension was met with different responses at the national level. While 'countries such as Albania and

2 Macedonia established diplomatic relations with the EU on 22 December 1995. For a chronological overview of EU–Macedonia relations, see Secretariat for European Affairs, Government of the Republic of Macedonia. Available at: http://sep.gov.mk/Default.aspx?ContentID=12.

Macedonia perceived the regional approach as a step backwards in their relations with Brussels, Bosnia-Herzegovina and Croatia saw in this EU undertaking an attempt at reconstituting somehow a version of former Yugoslavia' (Fakiolas and Tzifakis 2008: 381). In addition to its 'collective dimension' at the time of its launching, this initiative was considered deficient since it did not provide the basic incentive of the prospect of membership. Overall, the literature considered the regional approach to be inadequate to bring about stability and prosperity for not having a core perspective or an elaborate strategy, among other reasons (Fakiolas and Tzifakis 2008: 380–81).

Despite the critiques of its potential, the Regional Approach remains the first attempt of the EU to introduce formal conditionality in the Western Balkans and produced the first monitoring reports on compliance (EU Council 1997). Between 1997 and 2000 the Directorate for External Relations in the Commission published four reports on the fulfilment of the conditionality related to the Western Balkans, which can be considered as a general introduction to the EU's approach towards this region. In its first report on Macedonia, the Commission concluded that, while the country globally complied with democratic principles, the need for further progress was evident in relation to independent media, public administration reform, including the judiciary and law enforcement, higher education including for the Albanian minority, and so on (European Commission 1997). In this period, however, the reports of the European Commission were much less detailed than the Progress Reports in the post-2005 period examined below. Their shortcomings have already been noted in the literature in relation to the Eastern enlargement. Pridham argues that the first reports of the European Commission on the candidate countries lack a clear methodology on the political conditionality for objective cross-country comparisons between states, which explains the general assessment that the countries fulfilled the Copenhagen criteria (Pridham 2002a: 957). On the other hand, Kochenov has shown that despite the Commission's rhetoric concerning the 'absolute priority and importance of the political criteria, a record-low space in the progress reports is reserved for the political criteria analysis' (Kochenov 2004).

In the period studied, Macedonia was not at the top of the EU agenda and, compared to the Organization for Security and Cooperation in Europe (OSCE), the role of the EU pre-2000 was insignificant. Macedonia received attention in the early 1990s from scholars of conflict prevention and management looking at the other international organizations, not the EU (Ackermann 2000). Koinova highlights that analysis is missing because the EU had little involvement in Macedonia at the time (Koinova 2011: 810). Vachudova has questioned why the EU did not use political conditionality in the course of the 1990s since Macedonian governments were open to conditional Western assistance in this period (Vachudova 2003: 150). Stakeholders at the national level have explained this phenomenon against the background that many European representatives, especially the members of the European Parliament, believed that inter-ethnic relations in the country were completely relaxed, although they would later be subject to contestation.[3]

Overall, due to the weak engagement of the EU with Macedonia before 2001, it is difficult to classify its role under the headings explained in the theoretical section. This period, referred to as a 'pre-historic era' by my interlocutors, is characterized by the nascent development of the EU as a framework in the regional context and attempts at introducing the principle of conditionality. The approach of the EU in this period highlights the difficulties of employing conditionality without an objective of membership or a regional strategy, which has been confirmed in recent literature on the European Neighbourhood Policy (Schimmelfennig and Scholtz 2008). Building upon the experience of the Regional Approach, in the early 2000s the EU developed the SAA framework for the region, which introduced the perspective of membership examined in the following section.

2001–2004: From an Active Player to a Framework

The launching of the Stabilisation and Association Process followed the largely recognized failure of the Regional Approach in terms of its conditionality dimension. The biggest novelty of the SAP in comparison to its predecessor was the membership perspective for the region, which was confirmed on two occasions, in June and November 2000, at the Feira and Zagreb European Councils (EU Council 2000a, 2000b). The SAP consists of three main instruments supporting the countries from the Western Balkans to advance their development path towards membership: the SAAs, the autonomous trade measures and substantial financial assistance. The SAAs still represent the core formal contractual relationships between the countries of the Western Balkans and the EU. In addition, the SAA also established the main bodies of communication between the EU and the candidate countries: the Stabilisation and Association Council, the Committee and the thematic sub-committees. The SAP thus was based on the principles of the Eastern enlargement while at the same time attempting to address the post-conflict problems of this region. However, institutionally these countries were subject to monitoring by the Directorate General for External Relations. In other words, the region was still not part of the Directorate for Enlargement accession portfolio, which at the time consisted of the Eastern enlargement countries.

The SAP has largely been considered as 'a shift from an ad hoc "fire-fighting" style of crisis management to a more long-term broadly integrationist approach to the WB [Western Balkan] region' (Gordon et al. 2008). Macedonia signed the first SAA in the region in peculiar circumstances as the country was in the midst of an armed conflict. The significance of this first agreement has been highlighted as an implicit 'recognition of the leading role of Macedonia in the process of

3 Author's interview with former Vice Prime Minister for EU Affairs, Skopje, 23 December 2010.

stabilisation of the whole region' (Fierro 2003: 180). Against this background, analysts have argued that 'Macedonia was offered a SAA as a political reward for support given to the West during the Kosovo war' (Anastasakis and Bechev 2003). Moreover, 'being the first [Agreement] of this category to be concluded, the agreement sets up the precedent to follow' (Fierro 2003: 180). The SAA provided a framework for bringing the countries closer to the EU and its signing, among other issues, was politically conditioned as well. As a legal document, it does not contain a suspension clause, but linked the provision of assistance under Community Assistance for Reconstruction, Development and Stabilisation (CARDS) for the priorities set by the Commission, including democratic stabilization, strengthening civil society and the media, protecting minority rights and good governance (Börzel and Risse 2004).

Soon after signing the SAA, the country became subject to the regular monitoring on various aspects of the SAA, which included the political criteria. The Reports on the implementation of the SAA introduced in 2002 and issued until 2004 contain a thorough assessment of the political criteria for accession. Even though the SAA as an agreement does not cover the political criteria, the Commission has assessed the political situation and provided recommendations since 2001 when the Interim Agreement, which covered only trade and trade-related measures, entered into force (Calic 2003). Moreover, the political situation was regularly discussed in the meetings between the Republic of Macedonia and the EU in the SAA framework. Overall, the SAA reporting mechanisms were channels for stipulating political conditions and monitoring their realization, thereby formalizing political conditionality through the role of the EU as a framework for integration.

In addition to the SAA implementation, the EU involvement in Macedonia has been characterized by the increased activity for the signing and implementation of the Ohrid Framework Agreement (OFA) in 2001. During the 2001 crisis, the EU appeared as a key 'active player' in the conflict-prevention efforts through its role as a guarantor and even a co-signatory of the Agreement. The OFA, in turn, has become an essential element of EU political conditionality in the country. Whereas the implementation of the Agreement was channelled through domestic institutions, the EU constantly provided external support and at times pressure for its implementation.[4] Javier Solana, the EU High Representative for Common Foreign and Security Policy (CFSP), was directly involved in the OFA negotiations which led to the widely accepted maxim in the national political discourse that the road to Brussels leads through Ohrid. Not surprisingly, the implementation of the OFA has been monitored since 2002 in the Reports on the implementation of the SAA, the Progress Reports, the European Partnerships, as well as the SAA meetings. The linking of the implementation of the OFA to EU accession has mostly been evaluated as positive, since membership of the EU enjoyed high

4 Author's interview with former Vice Prime Minister for EU Affairs, Skopje, 23 December 2010.

support among the population in this period.[5] In other words, joining the EU gave the elites a common project that transcended ethnic divisions (Vachudova 2005b).

In addition to this form of political involvement, following the 2001 crisis, the first EU military peacekeeping mission Concordia was deployed in Macedonia, reinforcing the role of the EU as an active player in conflict prevention. The ensuing EU police mission Proxima was employed in April 2003 and was described by EU officials as 'an instrument in the cooperation between Macedonia and EU which needs to be used to meet the objective of membership in the Union' (Solana 2003). National stakeholders underlined that the EU had a major role on the ground with their police missions and as far as major policies were concerned.[6] As another interviewee put it, '[t]hrough the involvement on the ground, the EU gained a big stake in Macedonia's future. In some way it made Macedonia important to the EU almost rather than the other way around'.[7] This role was largely accepted by local elites as well. The president of Macedonia highlighted in a speech that 'our ambition is full membership in the Union, and I would like to see this mission … as a step in that direction. The more of the EU we have in Macedonia, the more of Macedonia there will be in the EU' (Trajkovski 2003).

The high involvement of the EU on the ground has also been reflected in the 'double hat' representation of the EU in the country, combining the position of the EU Special Representative of the Council and the Head of the Commission Delegation. According to one analyst, '[d]ue to this double hat policy, the political functions of the Council have been performed by the same representative managing issues of EU enlargement, further strengthening the link between the two'.[8] As a result, EU representatives in Macedonia have become intrinsically connected with domestic politics, which is demonstrated in the role they have played in forging consensus between the major political actors in the country. The very presence of the second hat of a special representative of the Council, which is foremost a political body, indicates that Macedonia is monitored and observed from Brussels in this segment as well.[9] This comprehensive type of EU conditionality has been described in the academic literature as 'a multidimensional and multi-purpose instrument, geared towards reconciliation, reconstruction and reform' (Anastasakis and Bechev 2003: 3).

The intense on-the-ground involvement of the EU between 2001 and 2004 was also characterized by increased administrative mobilization aimed at obtaining the status of a candidate country for EU membership, thereby facilitating the EU's role as a framework at the same time. In fact, the SAA implementation

5 Polls conducted in this period showed a steady support for EU membership ranging between 85 and 90 per cent.

6 Author's interview with former Vice Prime Minister for EU Affairs, Skopje, 23 December 2010.

7 Author's interview with think-tank analyst, Brussels, 11 October 2010.

8 Author's interview with think-tank analyst, Brussels, 11 October 2010.

9 Author's interview with former Ambassador to the EU, Tetovo, 22 December 2010.

was considered by the EU at this time as a condition for the country to apply for membership. In practice, however, this was not the case, because Macedonia applied for membership in 2004, indicating dissatisfaction with the status of a potential candidate. Hence, in 2004, the Government applied for EU membership and received from the Commission the Questionnaire on its membership application. In response, formal administrative structures for EU accession were established and, in fact, Macedonia was considered as a regional front-runner in the EU accession process in this period. A year later, the European Commission delivered a positive opinion on the country's membership. This opinion was upheld at the European Summit in December 2005, when Macedonia officially became a candidate country for EU accession. Croatia followed a similar path, illustrating that the countries in the region perceived the SAA as a way of deferring membership (Anastasakis and Bechev 2003).

Overall, in this period the EU became a major factor in the country through a combination of its engagement as an active player and as a framework for integration. Its active involvement on the ground was demonstrated through the OFA implementation, as well as the police missions deployed in the country. Rather than providing recommendations and standards, the Commission was at times engaged in assessing the domestic political scene and the relations between the domestic political actors. The literature links this attitude with the role the EU has acquired through the brokering of peace deals in Bosnia and Macedonia especially (Anastasakis and Bechev 2003). At the same time, the country also advanced through the regular mechanisms of EU integration by enabling the functioning of the joint bodies established under the SAA as well as obtaining candidate status. In line with the expectations of the theoretical literature, a weakening of the active role of the EU and facilitation of the framework dimension as a long-term phenomenon is to be expected in the post-2005 period examined below.

Post-2005: Combining the Roles of an Active Player and a Framework for Integration at an Advanced Stage of the Accession Process

Having applied for EU membership in 2004 and having obtained the status of candidate country for EU accession in 2005, Macedonia also became subject to EU conditionality through the regular progress reports and the European Partnerships. In fact, in 2005, in its relations with the region the EU introduced these formal instruments used in the previous enlargement, thereby developing its role as a framework for integration. These new instruments of monitoring were accompanied by institutional changes as the transfer of competences for the countries concerned from the Directorate General for External relations to the Directorate General for Enlargement. This change was 'a symbolic move testifying to the Union's strong commitment to the countries in question' (Bechev 2005). This set of reforms illustrates the post-2005 trend of developing the role of the EU in the region as a framework for integration in the long-term perspective. In other words, for the development and stabilization of this role, the Union needed to become a source of alternative solutions, rather than an actor involved in the everyday politics of the candidate countries (Noutcheva et al. 2004).

Contrary to these expectations, the latest enlargement with Bulgaria and Romania denoted a shift towards the increasing involvement of the EU in the political criteria of the acceding countries. In these two cases, the EU started moving down the path of high and low politics, areas traditionally regarded as internal to states (Pridham 2002b: 23). In the enlargement of 2004, research has shown that the Commission decided that it was against the logic of the accession process to monitor the political criteria again at the very end of the process (Sasse 2005: 10). Still, Bulgaria and Romania were subject to both extensive pre- and post-accession monitoring, especially in relation to the political criteria, illustrating an evolution of the Commission's approach to enlargement (Papadimitriou and Gateva 2009). A similar engagement of the EU can be traced in the Western Balkans in general, especially in the case of Macedonia. Vasilev is supportive of this role, stressing that 'when the country appeared to be straying from the desired course, the EU would move the government into action through formal statements of concern and criticism, coupled with clear guidelines on what they needed to do' (Vasilev 2011: 58).

Thus, the progress reports and the European Partnerships on Macedonia after 2005 still pay increasing attention to the fulfilment of the political criteria. This tendency is especially evident when compared to the progress reports and partnerships in the case of the new Member States before their accession, in terms of quantity and content. For example, the European Partnerships for the candidate countries of the Eastern enlargement traditionally contained fewer than five priorities in the political criteria, whereas in the case of the Western Balkans the partnerships contain approximately 30–50 priorities in the political criteria. The content and priorities of the partnerships are extremely important, not least because the Partnerships make the Copenhagen criteria legally enforceable since financial assistance from the Union is dependent upon meeting the partnership priorities (Kochenov 2004). Moreover, 'the Partnerships illustrate the extent to which the EU is seeking to influence the domestic and foreign affairs of membership candidates' (Smith 2003).

Two most pronounced areas where the Commission has been directly involved in the political sphere through the political criteria in Macedonia were the conduct of elections and political dialogue. In relation to the former, the 2008 European Partnership contained the following key priority for Macedonia: '[To] ensure that all future elections are conducted in accordance with the electoral code and deliver prompt decisions on any election irregularities and impose penalties that will deter further cases' (EU Council 2008). The activities aimed at fulfilling this criterion are elaborated in the national documents for European integration, indicating also the acceptance of EU's conditionality in relation to election issues (2008a). In relation to the second element, political dialogue, the 2008 European Partnership puts forward the following priority: '[To] promote a constructive and inclusive dialogue, in particular in areas which require consensus between all political parties, in the framework of the democratic institutions' (EU Council 2008). The rationale behind this condition, commonly referred to as 'political dialogue', was linked to the continuous boycott of the work of the Parliament by at least one major party between 2005 and 2007.[10] In fact, the lack of political dialogue was one of the main reasons for delaying the recommendation for starting the accession negotiations. This is clearly singled out in the 2008 Progress Report on Macedonia which states that 'as regards political dialogue, which is a key priority of the Accession Partnership significant efforts are needed' (European Commission 2008).

The EU's insistence on the notion of political dialogue has been met with differing responses in the academic and practitioners' circles. Vasilev argues that 'the EU was emphasizing the consensus principle in its rhetoric and feedback', hence the need for this condition (Vasilev 2011: 53–4). At the national level, nevertheless, the specific terminology of 'political dialogue' was linked to the further strengthening of political parties in the public sphere. The increased role of the parties has already been raised in both the academic and policy literature as one of the key challenges of the transformation process of Macedonia (Kačarska 2008). The lines of separation among the executive, legislative, and judicial bodies become blurred as one political party gains control (after elections) of both the legislature and the executive, which in turn provides an opportunity for partisan changes in the judiciary (Tsekov 2005). As a result, domestic analyses have emphasized that, rather than emphasizing political dialogue, the Commission should insist on 'policy dialogue' in Macedonia (Risteska 2010).

The extensive political conditionality of the EU towards Macedonia is evident not only through the documents, but has been clearly highlighted at several specific events in the post-2005 period. A notable example is the 2006 parliamentary elections, in which the EU Head of Delegation and Special Representative of the Council commented that 'it would be logical if the Government consisted of the parties that won the most votes' (Media Mirror 2006). At the national level, the statement was widely interpreted as an attempt by the EU to influence the formation of the government coalition, mostly with respect to the party representing the Albanian community. National stakeholders have singled out this event as an example of 'direct interference from the EU' and have generally considered the statement as a disruption of the internal dynamics of the Macedonian political scene.[11] In a similar vein, the position of the EU Special Representative has been considered controversial in itself. The literature has argued that 'the more advanced the transition process, and the more stable the security situation, the more relevant

10 In the period 2005–2007, one of the two major parties of the Albanian community with representatives in Parliament (Democratic Party of the Albanians and Democratic Union for Integration) was not participating in the work of this body.

11 Author's interview with former Vice Prime Minister for EU Affairs, Skopje, 25 December 2010.

the role of the Commission and the less central the position of the EU Special Representative' (Adebahr and Grevi 2007). However, in the case of Macedonia, this has not been the case. The continuous combination of the two roles resulted in a situation in which 'Macedonia is a candidate for EU membership and is being monitored from the Council. Hence, there is a dilemma whether the EU is supporting the country or is intervening with elements of a soft protectorate'.[12]

Hence, the EU is maintaining the features of an active player and a framework for integration although the country is at an advanced point of its accession process at which the political criteria for accession should have been fulfilled. This ongoing hands-on approach has instigated discussions within the EU institutions and at the national level. A related debate is ongoing within the European Commission concerning how long this involvement should last. A high-ranking Commission official pointed that 'the view of the Commission in general is that we should not perpetuate this policy. In relation to political criteria, a state must be sovereign. We have no business in deciding on directions'.[13] A similar dilemma regarding the role of the Commission is increasingly present at the national level as well. A former Vice Prime Minister for EU affairs in the country stressed that 'Macedonia has reached a point where the EU needs to take its hand of the saddle. The elites need to reach an agreement by themselves, instead of trying to communicate through an external actor'.[14]

Lastly, the issue that further adds complexity to the role of the EU in terms of evolving to a framework for integration is the dispute between Macedonia and Greece over the constitutional name of the country.[15] This dispute has been directly impeding Macedonia's accession process since 2009 when the Commission recommended the start of the accession negotiations dependent upon finding a mutually acceptable solution to the name issue. In its reports, the Commission underlines that 'maintaining good neighbourly relations, including a negotiated and mutually acceptable solution to the name issue, under the auspices of the UN, remains essential' (2009). The recommendation for negotiations has been confirmed every year since, but there is no progress on resolving this issue and therefore, negotiations have not commenced. At the national level, the name issue is considered as an interference with the expected course of conditionality, thereby weakening the credibility of the membership perspective. Since 2009, therefore,

12 Author's interview with former Ambassador to the EU, Tetovo, 22 December 2010.

13 Author's interview with European Commission official, Brussels, 15 November 2010.

14 Author's interview with former Vice Prime Minister for EU Affairs, Skopje, 25 December 2010.

15 Greece objects to the constitutional name 'Republic of Macedonia' and, as a result, the country joined the UN in 1993 under a provisional name 'The former Yugoslav Republic of Macedonia'. Both countries are engaged in UN-mediated talks to find a solution. However, even after a decade the talks have not produced results. For a factual background on the name dispute, see Karajkov (2008).

'the EU accession in the case of Macedonia consists of a lot of by-products without the main elements of the process of EU integration'.[16]

The dispute has added complexity to the intertwined roles of the EU in the domestic context and has added a further challenge to the evolution of the EU from an active player to a framework for integration. Commission representatives considered the issue as exogenous to the accession process, but it creates problems because it impedes upon the regular functioning of conditionality.[17] As a result, both of the roles of the EU as an active player and a framework are being compromised. Specifically with respect to the increasing interference of the EU as an active player in these conditions, national stakeholders have argued that 'the European representatives have to choose their battles carefully, rather than going into all of them because they will not be able to influence positively'.[18] With the ongoing name issue and without progress in the accession process, 'the increasing EU involvement in Macedonia has even created underlying resistance results in retreat to nationalism'.[19] As a result, the analysis of the Macedonian case shows that the combination of an increasing involvement in politicized issues largely confirms Sasse's findings on the potential polarization of deeper 'societal problems which might, in fact, have become ingrained in the context of the EU's involvement' (Sasse 2009: 28).

Conclusion

The analysis of the evolution of political conditionality since the late 1990s has illustrated the intertwined role of the EU as an active player and as a framework of integration through the example of Macedonia. The chapter examined the EU political conditionality as part of the Regional Approach of the late 1990s, the SAA process and in the post-2005 period. The intertwined roles of the EU were examined focusing on conditionality as a mechanism of Europeanization in the candidate countries. Supporting arguments for a process-based definition of conditionality encompassing its formal and informal features, the chapter also stressed the potential of this mechanism in contributing to societal polarization. Due to the specificities of the post-conflict nature of the Macedonian case, the chapter operated with the distinction between the EU as an 'active player' in conflict resolution and as a 'framework for integration', as developed by Noutcheva et al. As an active player, the EU gears the involved parties with short-term strategies and provides them with the external push leading towards conflict settlement. The

16 Author's interview with think-tank director, Skopje, 22 December 2010.

17 Author's interview with European Commission official, Brussels, 15 November 2010.

18 Author's interview with former Vice Prime Minister for EU Affairs, Skopje, 23 December 2010.

19 Author's interview with think-tank director, Skopje, 22 December 2010.

traditional instruments for accession, on the other hand, provide for the more long-term perspective of integration of Macedonia in the EU through the role of the EU as a framework (Noutcheva et al. 2004).

The empirical analysis was divided into three separate periods: the Regional Approach, the SAA period and the candidate status period post-2005. The chapter showed that, prior to 2001, there was weak involvement of the EU in Macedonia in terms of political conditionality. The analysis of the period between 2001 and 2004 illustrated the intertwined roles of the EU as an active player and as a framework for integration. The EU established its position as a significant player on the political landscape in the country with the signing and implementation of the OFA, as well as through its peacekeeping and police mission. At the same time, through its SAA implementation and application for membership, the EU also developed as a framework for integration. The combination of these two roles has continued in the post-2005 period especially in relation to electoral politics, as well as in relations between the major political actors in the country. The continuous involvement of the EU as an 'active player' on the domestic scene has nevertheless raised dilemmas among the Commission and at the national level, highlighting that the EU's involvement as an active player continued and was not a short-term phenomenon.

The chapter also demonstrated that the dispute over Macedonia's constitutional name further complicated the transition of the EU from an active player to a framework in the Macedonian context by questioning the legitimacy of the EU involvement. This status quo has stirred debate in the EU and national spheres, summarized by one of my interlocutors in the following way: 'Leaving Macedonia out of the enlargement process has had a negative impact on the interethnic relations, so it qualifies any EU success there.'[20] At the end, highlighting the potential for increased polarization and the problems of combining the two roles of the EU, this chapter also demonstrated the unintended impact of EU conditionality.

The conclusions of the presented analysis gain importance in light of the enlargement plans the European Commission announced in its 2011 Enlargement Strategy. In October 2011, the Commission pronounced that it will put even more emphasis on the political criteria for EU accession in the case of the Western Balkans. Hence, the first and the last chapter to be opened and closed in the upcoming accession negotiations with any of the countries in the region will be Chapter 23 – Judiciary and Fundamental Rights and Chapter 24 – Justice, Freedom and Security. These two chapters will be continuously monitored based on specific benchmarks and action plans (European Commission 2011). These plans indicate that the Commission intends to further intertwine the roles of the EU as an active player and a framework in the region. As a result, some of the caveats concerning the increasing EU engagement highlighted in this chapter carry regional significance, especially in light of the extensive role of the Commission in Bosnia and Kosovo, as well as the regular political stalemates in Albania.

20 Author's interview with think tank analyst, Brussels, 5 October 2010.

References

Ackermann, A. 2000. *Making Peace Prevail*. Syracuse, NY: Syracuse University Press.

Adebahr, C. and Grevi, G. 2007. The EU Special Representative: What Lessons for the EEAS? *EPC Working Paper*, No. 28, 56–64.

Anastasakis, O. 2008. The EU's Political Conditionality in the Western Balkans: Towards a more Pragmatic Approach. *Southeast European and Black Sea Studies*, 8(4), 365–77.

Anastasakis, O. and Bechev, D. 2003. EU Conditionality in South East Europe: Bringing Commitment to the Process. *South East European Studies Programme*. Available at: http://www.sant.ox.ac.uk/seesox/anastasakis_publications/EUconditionality.pdf [accessed: 15 March 2010].

Bartlett, W. and Samaržija, V. 2000. The Reconstruction of South East Europe, the Stability Pact and the Role of the EU: An Overview. *MOCT-MOST: Economic Policy in Transitional Economies*, 10(2), 245–63.

Bechev, D. 2005. EU and the Balkans: The Long and Winding Road to Membership. *South East European Studies Programme*. Available at: http://www.sant.ox.ac.uk/seesox/opinionpieces/bechev2005_2.pdf [accessed: 15 March 2010].

Berg, E. and van Meurs, W. 2002. Borders and Orders in Europe: Limits of Nation- and State-Building in Estonia, Macedonia and Moldova. *Journal of Communist Studies and Transition Politics*, 18(4), 51–74.

Bieber, F. 2011. Building Impossible States? State-Building Strategies and EU Membership in the Western Balkans. *Europe-Asia Studies*, 63(10), 1783–802.

Börzel, T.A. and Risse, T. 2004. One Size Fits All! EU Policies for the Promotion of Human Rights, Democracy and the Rule of Law. *Democracy Promotion Workshop*. Stanford University, CA: Center for Development, Democracy and the Rule of Law.

Bulmer, S. 2007. Theorizing Europeanization, in *Europeanization: New Research Agendas*, edited by P. Graziano and M.P. Vink. Basingstoke: Palgrave Macmillan, 46–58.

Calic, M.J. 2003. EU Policies towards the Balkans: Fostering Ownership of Reforms. *The International Spectator*, No. 3, 111–23.

Chandler, D. 2008. Normative Power and the Liberal Peace: A Rejoinder to John O'Brennan. *Global Society*, 22(4), 519–29.

EU Council 1997. Conclusions of the General Affairs Council. Brussels, 29 April 1997. Available at: http://www.consilium.europa.eu/ueDocs/cms_Data/docs/pressdata/en/gena/028a0057.htm [accessed: 7 February 2012].

EU Council 2000a. Presidency Conclusions. Santa Maria da Feira, 19–20 June 2000. Available at: http://www.consilium.europa.eu/uedocs/cms_data/docs/pressdata/en/ec/00200-r1.en0.htm [accessed: 7 February 2012].

EU Council 2000b. Zagreb Summit: Final Declaration. Zagreb, 24 November 2000. Available at: http://www.consilium.europa.eu/ueDocs/cms_Data/docs/pressdata/en/er/Declang4.doc.html [accessed: 7 February 2012].

EU Council 2008. *Council Decision on the Principles, Priorities and Conditions Contained in the Accession Partnership with Croatia and Repealing Decision 2006/145/EC*. Brussels, 12 February 2008. 2008/119/EC.

European Commission 1997. *Regional Approach to the Countries of South-Eastern Europe: Compliance with the Conditions in the Council Conclusions of 19 April 1997*. Brussels, 3 October 1997. Available at: http://ec.europa.eu/enlargement/pdf/key_documents/9710_report_a_en.pdf [accessed: 7 February 2012].

European Commission 2008. *The Former Yugoslav Republic of Macedonia 2008 Progress Report*. Brussels, 5 November 2008. SEC(2008)2695.

European Commission 2009. *The Former Yugoslav Republic of Macedonia 2008 Progress Report*. Brussels, 14 October 2009. SEC(2009)1335/3.

European Commission 2011. *Communication from the Commission to the European Parliament and the Council: Enlargement Strategy and Main Challenges 2011–2012*. Brussels, 12 October 2011. COM(2011)666 final.

European Commission 2012. Start of the High Level Accession Dialogue with the government of the former Yugoslav Republic of Macedonia. *Press Release*. Skopje, 15 March 2012. MEMO/12/187.

Fakiolas, E. and Tzifakis, N. 2008. Transformation or Accession? Reflecting on the EU's Strategy towards the Western Balkans. *European Foreign Affairs Review*, 13(3), 377–98.

Fierro, E. 2003. *The EU's Approach to Human Rights Conditionality in Practice*. The Hague: Martinus Nijhoff.

George, A.L. and Bennett, A. 2005. *Case Studies and Theory Development in the Social Sciences*. Cambridge, MA: The MIT Press.

Gordon, C., Sasse, G. and Sebastian, S. 2008. Specific Report on the EU Policies in the Stabilisation and Association Process. *MIRICO Special Report*. Bolzano: EURAC.

Government of the Republic of Macedonia 2008. *Blueprint on the Realisation of the Recommendations in the European Commission Progress Report on the Republic of Macedonia 2008*. Skopje: Vlada na Republika Makedonija.

Grabbe, H. 2006. *The EU's Transformative Power: Europeanization through Conditionality in Central and Eastern Europe*, Basingstoke: Palgrave Macmillan.

Héritier, A. 2005. Europeanization Research East and West: A Comprative Assessment, in *The Europeanization of Central and Eastern Europe*, edited by F. Schimmelfennig and U. Sedelmeier. Ithaca, NY: Cornell University Press, 199–209.

Hughes, J., Sasse, G. and Gordon, C. 2004. Conditionality and Compliance in the EU's Eastward Enlargement: Regional Policy and the Reform of Sub-national Government. *Journal of Common Market Studies*, 42(3), 523–51.

Hughes, J., Sasse, G. and Gordon, C. 2005. *Europeanization and Regionalization in the EU's Enlargement to Central and Eastern Europe: The Myth of Conditionality*. Basingstoke: Palgrave Macmillan.

Kačarska, S. 2008. Political Parties and the State in the Republic of Macedonia: Implications for Democratic Consolidation. *CEU Political Science Journal*, 3(1), 49–71.

Karajkov, R. 2008. Facts on the Macedonian-Greek Name Dispute. *Osservatorio Balcani e Caucaso*. Available on: http://old.osservatoriobalcani.org/index.php/article/articleview/8811/1/407/ [accessed: 20 June 2012].

Kochenov, D. 2004. Behind the Copenhagen Façade: The Meaning and Structure of the Copenhagen Political Criterion on Democracy and the Rule of Law. *European Integration online Papers*, 8(10). Available at: http://eiop.or.at/eiop/pdf/2004-010.pdf [accessed: 2 April 2012].

Koinova, M. 2011. Challenging Assumptions of the Enlargement Literature: The Impact of the EU on Human and Minority Rights in Macedonia. *Europe-Asia Studies*, 63(5), 807–32.

Magen, A. and Morlino, L. 2009. Hybrid Regimes, the Rule of Law and External Influence on Domestic Change, in *International Actors, Democratization and the Rule of Law: Anchoring Democracy?*, edited by A. Magen and L. Morlino. London: Routledge, 1–25.

Manners, I. 2002. Normative Power Europe: A Contradiction in Terms? *Journal of Common Market Studies*, 40(2), 235–58.

Media Mirror 2006. Parliamentary Elections 2006. Available at: www.nvoinfocentar.org.mk/_event_document.asp?key=57 [accessed: 1 February 2012].

Noutcheva, G., Tocci, N., Coppieters, B., et al. 2004. Europeanization and Secessionist Conflicts: Concepts and Theories. *Journal on Ethnopolitics and Minority Issues in Europe*, 1, 1–35.

O'Brennan, J. 2008. The EU in the Western Balkans: Enlargement as Empire? A Response to David Chandler. *Global Society*, 22(4), 507–18.

Papadimitriou, D. and Gateva, E. 2009. Between Enlargement-Led Europeanisation and Balkan Exceptionalism: An Appraisal of Bulgaria's and Romania's Entry into the European Union. *Perspectives on European Politics and Society*, 10(2), 152–66.

Pridham, G. 2002a. EU Enlargement and Consolidating Democracy in Post-Communist States: Formality and Reality. *Journal of Common Market Studies*, 40(3), 953–73.

Pridham, G. 2002b. The European Union's Democratic Conditionality and Domestic Politics in Slovakia: The Mečiar and Dzurinda Governments Compared. *Europe-Asia Studies*, 54(2), 203–27.

Pridham, G. 2005. *Designing Democracy: EU Enlargement and Regime Change in Post-Communist Europe*. Basingstoke: Palgrave Macmillan.

Radaelli, C.M. 2000. Whither Europeanization? Concepts Stretching and Substantive Change. *European Integration online Papers*, 4(8). Available at: http://eiop.or.at/eiop/texte/2000-008.htm [accessed: 5 November 2011].

Risse, T., Green Cowels, M., Caporaso, J. 2001. Europeanization and Domestic Change: Introduction, in *Transforming Europe: Europeanization and Domestic*

Change, edited by M. Green Cowels, M., Caporaso and T. Risse. Ithaca, NY: Cornel University Press, 1–18.

Risteska, M. 2010. Politički dijalog ili dijalog na politiki? *Dnevnik*, 17 September 2010.

Sasse, G. 2005. EU Conditionality and Minority Rights: Translating the Copenhagen Criterion into Policy. *EUI Working Papers RSCAS No. 2005/16*. Available at: http://www.eui.eu/RSCAS/WP-Texts/05_16.pdf [accessed: 3 May 2011].

Sasse, G. 2009. Tracing the Construction and Effects of EU Conditionality, in *Minority Rights in Central and Eastern Europe*, edited by B. Rechel. London and New York, NY: Routledge, 17–31.

Schimmelfennig, F. and Scholtz, H. 2008. EU Democracy Promotion in the European Neighbourhood: Political Conditionality, Economic Development and Transnational Exchange. *European Union Politics*, 9(2), 187–215.

Schimmelfennig, F. and Sedelmeier, U. 2004. Governance by Conditionality: EU Rule Transfer to the Candidate Countries of Central and Eastern Europe. *Journal of European Public Policy*, 11(4), 661–79.

Schimmelfennig, F. and Sedelmeier, U. 2005. Introduction: Conceptualizing the Europeanization of Central and Eastern Europe, in *The Europeanization of Central and Eastern Europe*, edited by F. Schimmelfennig and U. Sedelmeier. Ithaca, NY: Cornell University Press, 1–28.

Sedelmeier, U. 2006. Europeanisation in New Member and Candidate States. *Living Reviews in European Governance*, 1(3). Available at: http://europeangovernance.livingreviews.org/Articles/lreg-2006-3/ [accessed: 3 February 2012].

Smith, K.E. 2001. Western Actors and the Promotion of Democracy, in *Democratic Consolidation in Eastern Europe*, edited by J. Zielonka and A. Pravda. Oxford: Oxford University Press, 31–57.

Smith, K.E. 2003. The Evolution and Application of EU Membership Conditionality, in *The Enlargement of the European Union*, edited by M. Cremona. Oxford: Oxford University Press, 105–40.

Solana, J. 2003. Remarks at the Ceremony Marking the Ending of NATO's Operation Allied Harmony and Transfer of Authority to the EU. Skopje, 31 March 2003. Available at: http://www.nato.int/docu/speech/2003/s030331b.htm [accessed: 15 March 2012].

Trajkovski, B. 2003. Speech at the Ceremony Marking the Ending of NATO's Operation Allied Harmony and Transfer of Authority to the EU. Skopje, 31 March 2003. Available at: http://www.nato.int/docu/speech/2003/s030331d.htm [accessed: 15 March 2012].

Tsekov, G. 2005. Nations in Transit: Macedonia. *Freedom House*. Available at: http://www.freedomhouse.org/report/nations-transit/2005/macedonia [accessed: 10 January 2012].

Vachudova, M.A. 2003. Strategies for Democratization and European Integration in the Balkans, in *The Enlargement of the European Union*, edited by M. Cremona. Oxford: Oxford University Press, 141–60.

Vachudova, M.A. 2005. Promoting Political Change and Economic Revitalization in the Western Balkans: The Role of the European Union. *Slovak Foreign Policy Affairs*, 6(2), 67–73.

Vasilev, G. 2011. EU Conditionality and Ethnic Coexistence in the Balkans: Macedonia and Bosnia in a Comparative Perspective. *Ethnopolitics*, 10(1), 51–76.

Vink, M.P. and P. Graziano. 2007. Challenges of a New Research Agenda, in *Europeanization: New Research Agendas*, edited by P. Graziano and M.P. Vink. Basingstoke: Palgrave Macmillan, 3–20.

Chapter 5
Montenegro and the EU: Changing Contexts, Changing Roles

Jelena Džankić

This chapter looks at the transformation of the role of the European Union (EU) in Montenegro. It argues that the changing political context in the region induced shifts in the EU's approach to the smallest of the post-Yugoslav states. In supporting this argument, the chapter first looks at the EU's approach to Montenegro in the first years after the disintegration of Yugoslavia, when Montenegro was a constituent state of the Federal Republic of Yugoslavia (FRY) and when the ruling Montenegrin elites were associated with Slobodan Milošević. The chapter further argues that the first significant relational shift between Montenegro and the EU occurred in 1997, when the ruling Democratic Party of Socialists (DPS) split in two. The fact that the DPS faction, which remained in power in Montenegro, opposed the regime in Belgrade induced a more favourable approach from the EU towards the republic. A further change in the relationship between Montenegro and the EU occurred in the context of the divide over statehood and identity after Milošević was ousted from power in Serbia. By looking at the process of transformation of the FRY into the State Union of Serbia and Montenegro, the chapter highlights the development of the EU's conditionality in the Western Balkans. This is supplemented by an analysis of the EU's role in the post-independence period, whereby the major political compromises in Montenegro have emerged as a result of the aspiration to join the EU.

Introduction

At the time of the dismantlement of the Iron Curtain, the European Community (EC) was marching at an accelerated pace in the attempt to construct an institutionalized mechanism for European Political Cooperation (EPC) and assert itself as an important factor in global affairs. However, the collapse of Yugoslavia resulted in a series of difficulties which undermined the reputation of the EC and later on the European Union. In terms of effectively resolving the crisis, 'the civilian power Europe' failed to turn its resources into effective power and provide a framework for peace-making, resulting in the perpetuation of its dependence on the United States' (US) political and military resources. Yet, although in the 1990s the EU Member States failed to act cohesively, in recent years, the Common

Foreign and Security Policy (CFSP) of the EU has proven to be a viable tool for political cooperation among the Member States. This internal cooperation also strengthened the EU's enlargement mechanism, which thus manifested itself as a positive incentive for the democratization efforts of countries in the Balkans. As a consequence, the process of Europeanization, mirrored predominantly in the EU's conditionality, has emerged as the most efficient driver of domestic change.

The effects of this normative and substantial change in the EU's approach to the post-Yugoslav states were also tangible in Montenegro, the smallest of the former Yugoslav republics. At the time of the Yugoslav disintegration, Montenegro was the only one of the former Yugoslav republics that did not seek independence. At the referendum held on 1 March 1992, 95.4 per cent of the 66 per cent turnout expressed their preference for Montenegro to remain in a common state with other former Yugoslav republics wishing to do so (International Crisis Group 2000: 6). As a consequence of the referendum, Montenegro entered a common state with Serbia – the Federal Republic of Yugoslavia – in 1992. Hence, Montenegro sided with Serbia in the political course that the FRY took in the first half of the 1990s, which included involvement in conflict in Croatia and Bosnia and Herzegovina, international embargo, and general economic and societal downfall. In 1997, however, the political circumstances in Montenegro changed, marking a new era in the republic's relationship with the EU. The Montenegrin ruling party of reformed communists – the Democratic Party of Socialists – that captured most of the popular support split in two. One faction was led by the then President of Montenegro, Momir Bulatović, while the other was headed by the Prime Minister, Milo Djukanović. Initially, the conflict stood for either continuation of support for Slobodan Milošević – endorsed by Momir Bulatović's wing of the DPS – or opposition to the regime in Belgrade – advanced by Milo Djukanović. In order to obtain external political and financial assistance for countering the Belgrade regime, Djukanović's wing of the DPS turned to the West. As this faction of the DPS remained in power after the split, the government pushed a policy of 'creeping independence' (Roberts 2002: 6). The outcome of this policy was Montenegro's gradual estrangement from the Yugoslav federal institutions.

After the demise of Milošević in 2000, the internal Montenegrin debate was channelled through amplified demands for independence on behalf of Djukanović's government, and increased interaction of the opposition with the new Serbian elites to preserve the union. During this period, the key issue on the EU's agenda was to preserve peace and stability in the Balkans, and Javier Solana, the then High Representative for the CFSP, brokered an agreement between the Serbian and Montenegrin elites in 2002, thus transforming the FRY into Serbia and Montenegro in the following year. The final political contest between the pro-independence and the pro-union camps – the 2006 referendum on independence – occurred in a framework established by the international community, above all the EU. The EU's final guidelines for the referendum envisaged a lower limit of 55 per cent of the population voting for the question (minimum turnout of 50 per cent) as a condition for achieving the independence of Montenegro (International Crisis

Group 2005: 2). Following the peaceful separation of Serbia and Montenegro in June 2006, Montenegro emphasized EU integration as one of its political goals, and formally entered the accession process. It gained the status of candidate country in December 2010, with the date for opening of the negotiations still pending at the time of writing.

In order to prove its arguments, this chapter combines the theory on European foreign policy with the vast literature on Europeanization, supplementing it with empirical data acquired through the employment of qualitative methodology (documentary survey and context analysis). In the context of the evolution of the EU's foreign policy, the primary theoretical framework for this chapter is the 'capability-expectations' gap model, defined as the 'significant difference which had come about between the myriad hopes for and demands of the EU as an international actor, and its relatively limited ability to deliver' (Hill 1998: 3). Yet, in looking at the EU's capacity to deliver as an international actor, a multilayered concept of Europeanization is essential (Caporaso and Risse 2001, Goetz 2000, Grabbe 2001, Green Cowles, Hughes, Sasse and Gordon 2001, Knill and Lehmkuhl 1999, Lippert, Umbach and Wessels 2001). This does not merely relate to the political dynamics within the EU (Börzel and Risse 2000), but also to the normative effects of this process in the societies faced with ongoing institutional change (Hughes et al. 2004, Noutcheva and Emerson 2004, Schimmelfennig and Sedelmeier 2005). It is entrenched in the EU's political conditionality, not merely through the direct relationship with the aspiring members promulgated in contractual obligations and Progress Reports but also in the indirect effect of the Council of Europe and other international organizations, which also drive the EU integration process. As such, Europeanization is an essential conceptual tool for understanding the evolution of the approach of the EU towards the changing political milieu in Montenegro in the past two decades.

The EU and Montenegro, 1991–1997

In the last decades of the twentieth century, the global political scene was faced with a myriad of challenges. The collapse of communism moved the 'Montenegrin pendulum from one nexus of power to another', with major consequences for the republic as it was absorbed into the structures of the last Yugoslavia (Radonjić 1998: 25). In line with the recommendations of the European Community Arbitration Committee for the former Yugoslavia (the Badinter Commission), Montenegro was granted the right to self-determination along with the other Yugoslav republics. Consequently, the referendum of 1 March 1992 was aimed at resolving its status, since Yugoslavia was 'in the process of dissolution' (Rancharan 1997: 1261). At the referendum, 95.4 per cent of the 66 per cent turnout (minorities and the pro-independence Liberal Alliance of Montenegro boycotted the referendum) voted for the preservation of the common state (International Crisis Group 2000: 6). The outcome of the plebiscite ultimately prompted the adoption of the Constitution of the Federal Republic of Yugoslavia on 27 April 1992.

Furthermore, the political scene of Montenegro during the wars of Yugoslav disintegration was marked by the dominance of the Communist Alliance of Montenegro (DPS since 1990). The then unified DPS was allied with Milošević, and in relating to Montenegro, the EC/EU did not differentiate between the republics of the FRY. The collapse of Yugoslavia involved both the Montenegrin government and a vast proportion of the population in support of the operations of the Yugoslav Peoples' Army (JNA) in the territory of Croatia in 1991. Influenced by the media in perceiving the need to 'liberate Dubrovnik from the *ustaše* [the Croatian nationalist movement from the Second World War]' (Obala Production 2004), Montenegrin soldiers in the JNA attempted to revive the myth of heroism (Bieber 2003: 16).

Thus, the policy of the EC/EU towards Montenegro needs to be viewed within the context of the their overall approach to the disintegration of and the wars in the former Yugoslavia. In effect, in the aftermath of the Cold War, both internal and external expectations from the EC/EU were increased (Poos cited in Kinitis 1993: 148). In such an environment, the EC reactions to the collapse of Yugoslavia undermined the reputation of the Twelve as a serious global political actor. The disintegration of Yugoslavia showed the lack of cohesiveness among the Member States at the outbreak of the conflict.

Initially, the disintegration of Yugoslavia caused the Community a number of problems, since the Twelve failed to realize the dynamics of disintegration, and attempted to preserve the former Yugoslav federation at the beginning of the 1990s. However, upon the declarations of independence, first by Slovenia, followed by Croatia, the dissolution of the country proved to be a *malum discordiae* for the EC, which was too feeble to cope with a major crisis. Despite the attempts to coordinate their actions through the Troika of Foreign Ministers, which in mid-1991 concluded the agreement among the belligerents at the island of Brioni, The Hague conference in which the Community attempted to draft a solution to the conflict, and the arbitration attempts of the Badinter Commission, the divergence amongst Community members in terms of recognition proved the weakness of the EC/EU to deal with a major crisis in its surroundings.

In June 1991, the European Council was faced with the question of whether national aid to Yugoslavia should be suspended in the light of disintegration. Similar to the issue of recognition, the Member States of the EC were divided: Germany and the UK were in favour of suspension, Italy and France against, which resulted in failure to adopt the proposal by the Heads of Government. Furthermore, the question of embargo had geopolitical implications, since the disintegration of Yugoslavia would lead to the isolation of Greece, and deterioration of the country's economic position. The aforementioned divisions, therefore, evoked a lack of coherence in actions by the Member States, which reduced the Community's capability to act as a confident global actor, thus creating a myriad of problems for the EC, and subsequently the EU.

The inability of the EC at the beginning of the conflict to reach a common stance on the action to be taken in Yugoslavia proved that the EC at the time did not have the institutional resources to assert its 'actorness' on the global political scene (Edwards 1997: 173). This was clear even at the early stages of Yugoslav disintegration, and increased the reliance of the EC upon the UN since 1991 (Nuttall 1993: 12). The EC's position was exacerbated by the internal lack of coordination in terms of sending a military force to stop the belligerents, due to the opposition of the UK, and, more importantly, the lack of actual force to be sent in Yugoslavia. Furthermore, the events in Yugoslavia occurred at 'the hour of Europe' (Poos cited in Kinitis 1993: 148), that is, during the negotiations of the Maastricht Treaty, which increased both internal and external expectations of the Twelve to act at least coherently. In terms of internal expectations, the negotiations on the Treaty were perceived as a benchmark for the establishment of an ameliorated cooperative structure among the EC Members, which, accompanied by negotiations on economic and monetary union, raised expectations within the Community itself of being able to tackle an international crisis.

Yet the consequences of the EC/EU's policy in Montenegro were not felt only in the fact that the Community failed to prevent the Yugoslav breakup. Rather, the major socioeconomic effects were sparked by the EC/EU's imposition of sanctions on the FRY, of which Montenegro was a constituent republic. On 5 July 1991, as a consequence of the FRY's involvement in military operations in Croatia and Bosnia and Herzegovina, the EC imposed an arms embargo, while freezing all trade and financial aid to the ex-Yugoslav republics. This was furthered by subsequent EC trade sanctions, imposed against the former Yugoslav republics on 8 November 1991, and lifted a month later against all republics apart from Serbia and Montenegro. The two republics became sanction-free only after the conclusion of the 1995 Dayton–Paris Agreement, which ended the war in Bosnia and Herzegovina.

The general economic downfall was not the only result of these measures for Montenegro. The international embargo also allowed for the flourishing of a shadow economy, in which the DPS leadership played an important role (Uzelac 2003). According to Uzelac, the grip that the political elites held over the Montenegrin institutional set-up at the time of international embargo allowed for the creation of irregularities in the process of transition (Malešević and Uzelac 2007, Uzelac 2003: 105–6). These transitional irregularities allowed a small oligarchy to use the power vacuum at the top of society, created by the fall of the previous system, in order to seize the state and gain wealth and power (Medojević 2001). As will be explained in detail in the following section, the combination of these factors contributed to the perpetuation of the DPS's power in Montenegro, but it also had an important effect on the polarization of Montenegrin political and social life in 1997, which also changed the EU's policy towards the smaller partner in the FRY.

The EU and Montenegro, 1997–2000

The aftermath of the conflict period in the Balkans brought about a shift in the EU's approach to the common foreign policy, which increasingly became the 'silent disciplining power on the "near abroad"' (Wæver 1998: 99). Still, unlike the hard military power and coercive mechanisms ensuring compliance often used by the US throughout the 1990s, European foreign policy primarily involved the export of EU norms and values as a means of stabilizing the fragile Western Balkan region. This process was initiated immediately after the signature of the Dayton–Paris Agreement ending the conflict in Bosnia and Herzegovina, but did not gain momentum until the post-Nice period. The contextualization of the EU's approach to the region was of particular significance in Montenegro, which, after 1997, embarked on a distinct political course marked not only by domestic political polarization but also by detachment from the FRY institutions. The latter induced Montenegro to turn to the West (the US and the EU) for financial and political support to counter Milošević's regime. Although the US's support was manifestly larger during this period, the EU's striving to redeem its failures from the early 1990s allowed for a distinct EU policy towards Montenegro at the time when it was still a part of the federation with Serbia. The understanding of these complex and interrelated dynamics requires an overview of both the developments in the EU and those in Montenegro.

The post-Dayton approach of the EU involved the strengthening of regional initiatives with the countries for which the Council had not at that time developed guidelines for association agreements. Consequently, the political commitment by the countries to engage in promoting stability and growth in South Eastern Europe (SEE) was not established as an instrument of the EU's foreign policy, but through the Stability Pact for SEE (1999), further implemented by the Organization for Security and Cooperation in Europe (OSCE). This means that, although there was a certain degree of coherence in the European foreign policy at the time, it was the lack of capabilities for dealing with large-scale conflicts in a fragile political environment that prevented the redemption of the earlier failures of the EU related to the disintegration of Yugoslavia. Still, some advancement was brought about by the institutional changes at the EU level introduced by the Treaty of Amsterdam (entered into force in 1999), which represented a significant step in the legal codification of European values – progressively evolved to epitomize the 'Respect for Human Dignity, Freedom, Democracy, Equality, and the Rule of Law and respect for Human Rights, including the rights of persons belonging to minorities' (Roche 2007). Although most of these values had already been spelled out in the Maastricht Treaty, they were reiterated in 1993 in the Copenhagen criteria for accession, the Amsterdam Treaty, and at the Nice European Summit where the EU Charter of Fundamental Rights was adopted in view of a harmonized approach to the respective values. As a result of the development dynamics within the EU following the Nice Treaty, there was an intensification of the Union's involvement in the Western Balkans, yet through a strategy that involved the export of European values both by presence on the ground, stabilization initiatives and socialization into Europe. This reveals the progressive nature of the EU's 'redemption' in the region.

As noted in the introduction, in 1997 the Montenegrin ruling party of reformed communists (DPS) split in two. The political parties thus created formed the two axes of political attraction in Montenegro in the decade following the split. In the political discourse, the power *mêlée* created by the split originally revolved around endorsing or countering Milošević's regime. The wing of the DPS led by Milo Djukanović constructed its identity in opposition to the regime in Belgrade. The competing wing (Socialist People's Party as of 1998), led by Momir Bulatović, took on the role of supporter of Milošević's politics. Simultaneously, it emphasized the corrupt nature of Djukanović's DPS, and claimed that the Montenegrin ruling elites sought to retain their grip over the state that was created in the years of the international community's sanctions.

The resources of Djukanović's DPS, which remained in power in Montenegro, were largely dependent on external support in the years immediately following the split. In the period from 1997 to 2000, both the internal and the external support of Djukanović was built around his opposition to Milošević. By countering Milošević's policies in this way, the Montenegrin leadership embarked on a course of 'creeping independence'. The table below presents the gradual detachment of Montenegro from the FRY. As indicated in Table 5.1, many of the policies adopted by the Montenegrin government in the second half of the 1990s were responses to actions of the authorities in Belgrade. This shows that the policy of 'creeping independence', and thus the externalization of Belgrade by Djukanović's camp, were largely triggered by the latter's opposition to Milošević's politics.

By 2000, only a few competences remained at the federal level. Out of the competences that formally remained at the level of the common state, some were also partly enforced by Djukanović in Montenegro alone. Montenegro introduced several economic, fiscal and monetary policies that distanced it from the Yugoslav authorities, including customs and the Deutschmark. In addition, the DPS-SDP-NS government took control of the civil airports in Montenegro, as well as the railway services, thus transferring most of the transport and traffic regulation from the federal level to that of the republic. From 1997 until 2000, Montenegro, which had its own Foreign Ministry, opened representative offices in Washington (DC), New York, Brussels, Rome, London, Ljubljana and Sarajevo. The Citizenship Law of October 1999 instituted a separate category of Montenegrin citizenship and established a separate visa regime (Lukić 2005: 64–5). Additionally, in light of Djukanović's opposition to Milošević's politics, Montenegro was granted 'early beneficiary' status in the Stability Pact for South Eastern Europe (European Stability Initiative 2001: 7–9).

The by-product of such a policy, which entailed detachment from the federal institutions, was that the DPS gradually transformed its opposition to the regime in Belgrade into a quest for statehood. Wim van Meurs claims that the '[p]roponents of Montenegro's independence consistently painted a black-and-white picture of

Table 5.1 'Creeping independence' in practice by 2001

FRY action	Montenegrin reaction
1998 – the FRY authorities cease to pay Pension Contributions	1998 – after the FRY action, Montenegro diverted the payments to the Federal budget into an account of its own
1999 – FRY introduces a set of federal laws restricting imports for Montenegro	August 1999 – Montenegro takes over the Customs Administration
March–November 1999 – The National Bank of Yugoslavia prints currency in order to cover the expenses caused by the Kosovo crisis in 1999	November 1999 – Montenegro introduces the Deutschmark as a *parallel currency* November 1999 – Montenegro establishes the Montenegrin National Bank by taking over the Podgorica Branch of the National Bank of Yugoslavia November 1999 – Montenegrin Payments Bureau (ZOP) becomes autonomous November 2000 – the Deutschmark becomes the only *official currency* in Montenegro
1998–2000 – Milošević increases the presence of the VJ on Montenegrin soil, and exerts further political pressure on Montenegro and Kosovo, finally resulting in the NATO intervention	1998–2000 – the Montenegrin Ministry of Interior builds up a militarized police force of about 15,000 men, which would be able to counter VJ in the case of an attack 1999 – the Montenegrin government takes over the management of the civil airports in Podgorica and Tivat
1996–2000 – the FRY conducts isolationist foreign policy	1999 – Montenegro institutes a separate visa regime October 1999 – Montenegro institutes a special category of Montenegrin citizenship 2000 – Montenegro granted the status of 'early beneficiary' in the Stability Pact for SEE By 2001 – Montenegrin Foreign representation offices were established (mostly as trade Missions) in Washington (DC), Ljubljana, New York, Brussels, Rome, London and Sarajevo
By 2001 – most ties of the Montenegrin government with the FRY are broken; the regime in Belgrade changes	January 2001 – Montenegro allows the opening of a representative FRY office in Podgorica

Source: European Stability Initiative 2000, van Meurs 2003.

FRY as the state associated with the reactionary and repressive Milošević regime and of Montenegro as a paradise of pro-European reforms' (van Meurs 2003: 65). This rhetoric was further affirmed by international assistance, half of which came from the EU. The importance of sustaining the DPS's policy through financial assistance in the late 1990s was also highlighted by the European Stability Initiative (ESI) and International Crisis Group (ICG) analysts as an important aspect of Montenegrin 'Westernization' in the second half of the 1990s (European Stability Initiative 1999, International Crisis Group 1999, International Crisis Group 2000).

Table 5.2 EU financial assistance to Montenegro in million DM: 1999–2001

Category	Description	Million DM
Humanitarian aid	ECHO	**68.6**
Infrastructure	Obnova/CARDS, etc.	**97.0**
Democracy aid	EU – media project under Obnova	**3.0**
	special initiatives	**78.4**
Financial support	agricultural development, food programmes	**61.5**
	Obnova – refugees assistance	**25.5**
Total		**334. 0**

Source: European Stability Initiative 2001.

In the context of the data presented in Table 5.2, most of the EU's assistance to Montenegro was granted in 1998 and 1999, at the time of the crisis in Kosovo. In fact, another outcome of the DPS's government orientation towards the West was the acceptance of refugees from Kosovo and neutrality during the crisis. During 1998 and 1999, approximately 29,000 refugees and internally displaced persons from Kosovo, mostly of Albanian or Roma origin, found shelter in the territory of Montenegro, as well as some members of Serbia's pro-Western elites (European Council on Refugees and Exiles 2001: 151). The population of Montenegro increased by 15 per cent, which required special attention on the behalf of the EU. Hence the financial and political support thus provided to Montenegro is understandable in the framework of the Union's role as a stability factor in the post-Yugoslav space after the Dayton agreement. Yet, such support also reaffirmed the detachment of Montenegro from the FRY institutions, and induced a pro-independence drive on the behalf of the ruling elites. Given the complex situation in Kosovo, and the perception of the EU that the preservation of the common state would be a stabilizing factor for the fragile region, there was an inevitable shift in the EU's policy towards Montenegro after Milošević was ousted from power in Belgrade. These dynamics are described in the following section.

The EU and Montenegro, 2000–2006

Since 1999, most of the stabilization initiatives in the Western Balkans have been realized under the aegis of the Stabilisation and Association Process (SAP), aimed at assisting the Western Balkan countries in meeting the criteria of EU membership. The conditionality of European integration envisaged in SAP introduced a paradigm for domestic institutional reform, through its pressure on the post-Yugoslav countries. The SAP envisaged the political, financial, logistical and professional assistance through trade liberalization, financial assistance and the signature of the Stabilisation and Association Agreements (SAA). Hence, the export of European values through the conditionality of the EU accession process has largely been perceived as an impetus for internal institutional reform in the countries of the Western Balkans.

By the same token, with the change of the political landscape in the FRY after the fall of Milošević, the international actors acquired different roles. Instead of the US, the Balkan region was faced with the salient influence of the EU. The poor track record of the CFSP in the Balkans in the years of Yugoslav disintegration, accompanied by the fear of the domino effect, provided an incentive for the EU to redeem itself (Edwards 1997). However, the policies of the EU, and in particular the European Commission, at the time of the existence of the State Union proved to be rather inconsistent with striving to preserve the common state. As a consequence, the EU's involvement in Serbia and Montenegro from 2000 to 2006 provided an indirect support to the discord between the two components of the Union, while reinforcing the internal debate over statehood and identity in Montenegro. This was done, albeit in different ways, through the efforts of the High Representative for the CFSP, Javier Solana, to preserve the common state, and through the 'twin-track' approach to institutional reform in Serbia and Montenegro.

The CFSP and the Montenegrin Divide

The international involvement headed by the EU resulted in the signing of the Belgrade Agreement on 14 March 2002. The Belgrade Agreement, which essentially maintained the status quo between the two republics, outlined that in the case of Montenegrin secession 'Serbia would be the successor state, explicitly so concerning the implementation of the UNSC Resolution 1244 for Kosovo' (Government of the Republic of Montenegro 2002). During the negotiations between Serbia and Montenegro on the nature of the new state, Solana exerted considerable pressure on the Montenegrin government to postpone its plans for an early referendum. Given Djukanović's perceived image as 'Western-oriented' among the population of Montenegro, his political route was bound to follow the EU's guidelines. That, however, implied the delay of the independence project that he had embarked upon after the fall of Milošević. Van Meurs indicates that the brokering of the Belgrade Agreement was not only attributable to Solana's 'diplomatic arm-twisting', but also to the realization of the political actors 'that they had outmanoeuvred themselves (and others) into a "lose-lose" situation' (van Meurs 2003: 64).

Yet after the mediation of the Belgrade Agreement, the adoption of the Constitutional Charter of Serbia and Montenegro remained in a political deadlock for over a year and was ultimately resolved by another of Solana's interventions. The discord emerged with respect to the election of delegates to the assembly of the common state. Djukanović and his political allies advocated an appointment procedure, whereas the pro-union opposition rejected any proposals which did not contain provisions for direct elections (Phillips 2001). Another claim pushed the pro-union bloc to hold strongly to the idea of direct elections. The unionists enjoyed support from the Serbian authorities in their vision of the common state as a more centralized unit (*The Economist* 2004). Late in 2002, Solana fostered a resolution to the crisis, incorporated in the Constitutional Charter. The chief EU negotiator based his proposal on Djindjić's previous plan to have a transitory period of indirect elections, while the popular vote would take place only after two years (European Stability Initiative 2005, Jovanović 2005).

The Constitutional Charter was finally adopted in December 2002, while the law on its implementation followed two months later, since the minor coalition partners of both blocs in Montenegro, as well as the minor parties of the government in Serbia, did not favour the document because they believed that it lacked substance (Šuković 2002). Given the fact that the Charter was negotiated with difficulty, and that it did not contain the genuine consent of the elites of Serbia and Montenegro, it did not provide a functional framework for the common state. Rather, it further emphasized the claims of the Montenegrin pro-independence bloc that any centralizing policy would result in Serbian control over Montenegro (*Vijesti* 2003). As a consequence, Djukanović's bloc 'resisted almost all attempts at forming an effective and functional central State Union government' (International Crisis Group 2005: 5), with the aim of reassuring their voters that the common state was of a temporary nature, and that the independence cause was ongoing.

Given the discord among the elites of Serbia and of Montenegro, a further political divide in 2005 caused the near collapse of the structures of the State Union, which further reinforced Dukanović's claims. By that time, the prospect of a referendum in Montenegro was widely announced as a certainty. The issues that hampered the adoption of the Constitutional Charter re-emerged with the impending expiry of the three-year moratorium (International Crisis Group 2005: 9–12). According to the provisions of the Constitutional Charter, the representatives to the Assembly of the State Union were to be substituted by directly elected members in March 2005. Previously, they were appointed by the governments of Serbia and of Montenegro.

However, the Montenegrin government was reluctant to hold direct elections due to the lack of popularity of the State Union, its pro-independence agenda, as well as the fear of loss of influence at the federal level since the pro-union opposition was more likely to motivate its voters to go to the polls. This caused a fierce reaction in Belgrade and was described by Koštunica as 'a rude violation of the Belgrade Agreement' (*B92* 2005). Subsequently, part of the Serbian intellectual elites called for the independence of Serbia and the removal of Svetozar Marović, a DPS member, from the post of the President of the State Union. A further Solana intervention reduced the tensions between Serbia and Montenegro in the short term. The two members of the State Union agreed to change the provisions of the Constitutional Charter and hold elections for the Union's Assembly concurrently with those of the republics' parliaments in 2006. However, with so many issues lingering in the political space between Serbia and Montenegro, and as many in Montenegro alone, the perpetuation of uncertainty widened the gap between the pro-independence and unionist followers, giving leverage to emphasize the statehood and identity debate.

The Commission's 'Twin Track' for a 'Twin State'

The major distinction between the actions of Solana and that of the European Commission in dealing with the State Union of Serbia and Montenegro was a matter of approach, rather than of substance. Solana based his negotiations with the Serbian and Montenegrin elites on the conflict-mediation approach, thus trying to find a common denominator for a political agreement that would enable the sustainability of the common state, without reopening 'Pandora's box of border adjustments throughout the region' by prompting the early departure of Kosovo (Baillot 2004: 14). The Commission asserted that its involvement in the functioning of the State Union 'was that of internal harmonisation aimed at furthering its EU integration' (Tocci 2004: 566). However, it was precisely this difference in approach, and the Commission's inability to balance out its policies at the level of the State Union with those in Serbia and in Montenegro that actually gave an indirect backing to Djukanović's independence drive.

The main objective of the Commission was the creation of effective and functioning state structures within Serbia and Montenegro, which would facilitate the SAP. In the light of the integration of the region into the EU, the Brussels officials believed that '[e]ven though the two countries have no common currency, central bank or a unified market', it was more practical to conduct negotiations with a single country (Baillot 2004: 14). Unlike in the case of Solana's intervention, where the lack of clarity gave enough scope for distinct interpretations of the actions of the elites, in the SAP (guided by the Commission) it was precisely the clear division of competences between the constituencies and the State Union that allowed the pro-independence and the pro-union camps to capitalize on the EU's policies.

The latter was even more apparent in the area of economic relations between Serbia and Montenegro, which eventually resulted in the adoption of the 'twin track' for EU accession negotiations. In fact, this approach was adopted after a stalemate in the SAP of Serbia and Montenegro. The Feasibility Study for the State Union's membership prospect was supposed to be initiated late in 2003, but was postponed until early in 2004 due to the problems in the institutional set-up of the common state. The 'twin track' was unique in terms of the Commission's approach to aspiring members, as it allowed Serbia and Montenegro a large degree of flexibility in the process of integration. It highlighted the division of competences between the republics, and the shared competences of the republics vis-à-vis the State Union in the light of European integration. The EU proposed separate tracks for negotiations of institutional arrangements (harmonization of economies with the EU standards) with each of the states, whereas regional cooperation and non-economic international obligations fell within the jurisdiction of the common state (European Commission 2005).

Such an approach, based on the asymmetry of the population and economic resources of the two constituencies, also envisaged the synchronization of the two economies, which proved to be a rather difficult task owing to the distinctiveness of the respective systems (Djurić 2004, Gros et al. 2004, Medjak 2004). Moreover, Serbia and Montenegro had rather diverging foreign trade, monetary and privatization policies, which somewhat disrupted the practical application of the 'twin track'. Both in Serbia and in Montenegro, this eventually increased the general dissatisfaction with the functioning of the State Union, which, given the impossibility of internal economic harmonization, became described as 'a virtual community that *de facto* does not exist' (Batić cited in *Glas javnosti* 2004).

The second component of the 'twin track' (the shared competences) was another point of discord for Serbia and Montenegro, which gave a further push to Djukanović's government's independence drive. It generated claims on behalf of the Montenegrin government that the Serbian government was hampering the process of EU integration, by breaking the conditionality principle through lack of cooperation with the Hague Tribunal for war crimes (European Stability Initiative 2005). At the Thessaloniki Summit of 2003, the EU envisaged the 'promise of EU membership [as] the basis for all EU conditionality in the region, from compliance with the Hague Tribunal to institutional reforms, from trade liberalisation to the unresolved strategic issues' (European Stability Initiative 2005: 2). The Thessaloniki Agenda resulted in the reiteration of the Union's 'determination to fully and effectively support the European perspective of the Western Balkan countries, which will become an integral part of the EU, once they meet the established criteria' (EU Council 2003).

In early 2006, the negotiations for the SAA of Serbia and Montenegro with the European Union were blocked because of Serbia's lack of cooperation with the Hague Tribunal. This reinforced the message of the Montenegrin government that Montenegro was 'a hostage of Serbian politics' (*B92* 2006), through which the Montenegrin government justified its pro-independence course to the people. According to international analysts, the pro-independence camp claimed that no war criminals had ever been sheltered by the Montenegrin state authorities, who emphasized their commitment to cooperation with the international authorities (International Crisis Group 2005).

In general terms, the adoption of the 'twin track' of the Commission provided for a temporary appeasement of tensions between the governments of Serbia and Montenegro during the preparation of the Feasibility Study for the prospect of EU membership in future. This was mainly because the governments of the constituents of the State Union could interpret the approach as beneficial to their cause. However, the actual effect of the 'twin track' on Montenegrin politics and society was somewhat different in practice, because the 'twin track' was inconsistent with the provisions of the Belgrade Agreement related to the harmonization of economies. It gave the government enough space to claim that it retained the elements of Montenegrin statehood in the process of EU integrations, through an emphasis on the division of competences (Djurović 2005, Medjak 2004, Teokarević 2003). It also gave an indirect push to the claims of Djukanović's camps that the process of EU integration was being slowed down because of Serbia's failure to cooperate with the International Criminal Tribunal for the former Yugoslavia (ICTY). At the same time, it allowed the opposition to claim that the single SAA affirmed the EU's commitment to keep Montenegro in a common state with Serbia, and that 'Montenegro cannot survive on its own' (*Dan* 2005: 2).

Implications and Referendum

The EU's efforts to preserve the stability in the Balkans effectively exacerbated the internal fissure in Montenegro in the period from 2000 to 2006. This led many international analysts to conclude that, regardless of the changes, the new state was the last stage of the disintegration of the former Yugoslavia (Lukić 2003: 93). This was confirmed by the announcement of the prospect of the referendum on Montenegrin independence in late 2005. In January 2006, both Djukanović and the Speaker of Parliament, Krivokapić, claimed that the reality of the processes the region was faced with demanded the re-establishment of Montenegrin statehood. In interviews with the French and the UK ambassadors, Krivokapić maintained that '[r]eaching a consensus on referendum is not a matter of time; rather it is a question of political readiness and willingness to accept the reality of the process, which has been growing in strength in Montenegro in the past fifteen years' (*Pobjeda* 2006: 3).

Due to the importance of the final outcome for both camps, establishing the rules and principles the two parties would accept was difficult, even for the external intermediary. For the government, whose agenda was the independence of Montenegro, victory in the popular vote was connected with its own political survival. As a result of the general atmosphere of utmost political mistrust, the political discourse in early 2006 revolved around finding 'a constructive and efficient dialogue, through which either an acceptable solution or the lowest common denominator would be reached' (*Vijesti* 2006). Thus, international mediation was essential in breaking the deadlock related to the May 2006 referendum on independence. Referendum rules were established by the EU. By introducing guidelines for an internal compromise, the EU attempted to provide for a consensus through which the statehood issue would be resolved in a manner acceptable to both parties.

Hence, the final political contest between the pro-independence and the pro-union camps – the 2006 referendum on independence – occurred in a framework established by the international community, above all the EU. The EU's final guidelines for the referendum envisaged a lower limit of 55 per cent of the population voting for the question (with a minimum turnout of 50 per cent) as a condition for Montenegro gaining independence (International Crisis Group 2005: 2). The adoption of such a formula not only indicates the degree of the divide in Montenegro, but also the importance of the EU's involvement in the referendum process as a means of redemption for its failure in the Balkans in the early 1990s. The post-referendum relationship between Montenegro and the EU is described in detail in the following section.

Post-Referendum Montenegro and the EU

With the resolution of the status issue in 2006, the relationship between the EU and Montenegro took up a distinct course, in which the EU's conditionality mechanisms have proven to be the major driver of domestic political and institutional change. In fact, the stipulation of reciprocal mechanisms between Montenegro and the EU has revealed the power of both the EU's conditionality and of broader forces in operation in the Western Balkans under the aegis of Europeanization. From the perspective of the aspiring members, the process of Europeanization primarily refers to the socialization into Europe through 'processes of (a) construction (b) diffusion and (c) institutionalization of formal and informal rules, procedures, policy paradigms, styles, "ways of doing things" and shared beliefs and norms which are first defined and consolidated in the making of EU decisions and then incorporated in the logic of domestic discourse, identities, political structures and public policies' (Radaelli 2000: 4).

Consequently, the impact of Europeanization on the Balkan countries is reflected in the tendency to absorb the socio-institutional values of Europe and make them an essential part of the institutions and cultures stained with communist and post-conflict legacies. This absorption of values and mechanisms of the EU takes place either through direct policy transfer envisaged in the process of conditionality, or, indirectly, through the activities of other institutions, such as the Council of Europe, that implicitly instil European norms and values in a society. This means that, through Europeanization, either direct or indirect, domestic political actors become increasingly subject to the policy making of the EU, which is gradually augmented with the approximation of the country's status in the accession process (Buller and Gamble 2002). As the countries under consideration become increasingly norm-takers, the 'actorness' of the EU is manifested through the appearance of a higher degree of coherence, which has a positive effect on conditionality itself. Hence, understanding the current EU dynamics in Montenegro requires an overview of both the direct and indirect mechanisms through which the EU seeks to secure norm compliance in an aspiring member.

Formal Aspects of the EU–Montenegro Relationship

In fact, the formal relationship between the EU and Montenegro, through which the EU's conditionality has been exercised, along with the mechanisms of Europeanization, was established on 12 June, when the EU Council recognized Montenegro as an independent sovereign state. This was followed by a negotiating mandate for an SAA with Montenegro late in July the same year. In addition, after the parliamentary elections in 2006, the SAA negotiations were launched, and the European Partnership for Montenegro followed in January 2007 (European Commission 2011). The official signing of the SAA took place in Luxembourg on 15 October 2007, a mere four days before the adoption of the Constitution of Montenegro. By signing the SAA, Montenegro formally committed to functionally implementing policies that would have as their final goal the approximation to the legal and political order of the EU. Yet, the document was not ratified by all the current 27 EU Member States until 1 May 2010, and in the period between the signature and the date of entry into force, the Interim Agreement on Trade and Trade-related Issues was operational between Montenegro and the EU. Nevertheless, the fact that the entry into force of the SAA was postponed for three years following the document's signature does not merely show the internal Montenegrin dynamics of adaptation to EU's rules. Rather, it also reveals the length of procedures in the EU, whereby ratification is required in all of the current Member states.

A further Montenegrin commitment to EU integration was stipulated on 15 December 2008, when the young European country submitted its formal application for EU membership (European Commission 2011). By late April the following year, the Council of Ministers recommended to the European Commission to submit the *avis* (opinion) on Montenegro's application for EU membership. As envisaged in the conditionality mechanism of the EU, the Commission's assessment was based on the country's answers to the Questionnaire. The Questionnaire, which the Montenegrin authorities responded to from July to December 2009, was based on requests for information on the capacities of the state to transpose the Union's *aquis*, but also to meet the political criteria for accession, including the rule of law, and protection of human and minority rights, as well as the capacities of Montenegro to deal with the competitive pressures that the small country's economy would be exposed to following its accession to the EU. Montenegro became an official candidate country for EU membership on 17 December 2010, while the Commission recommended the official start of negotiations in October 2011. The date on which the negotiations will be started depends on Montenegro's progress in dealing with its transitional problems (European Commission 2011).

Conditionality and Europeanization at Work

The relationship of Montenegro with the EU required more than the conclusion of a number of agreements. It was precisely in the underpinning transformation of Montenegro required in the light of EU accession that the EU's conditionality, and Europeanization in general, were at their best. The most prominent drivers of change in Montenegro were: (a) the visa liberalization process, which was the strongest conditionality mechanism that the EU has exercised in the Western Balkans; and (b) the recommendations stipulated in the EU's Opinion on Montenegro's preparedness for EU membership. The reason these two mechanisms of conditionality had the strongest effect is precisely because they required Montenegro to make a specific series of political changes that would yield a tangible outcome in terms of the country's relationship with the EU.

On 19 December 2009, Montenegrin citizens were granted visa-free travel to the Schengen area (European Commission 2011). Rather than being just the culmination of the Roadmap for Visa Liberalisation, this process also had a high symbolic value for Montenegrin citizens. The importance of visa-free travel in Montenegro is not only related to the practicalities of travel, rather, it has a historical and emotional dimension, since the citizens of the socialist Yugoslavia enjoyed visa-free travel both to the communist east and to the countries of the then European Economic Community. As a result of the series of violent conflicts, the citizens of the successor states of the former Yugoslavia lost their privilege of visa-free travel. The subsequent initiation of the SAP in 1999 offered these countries the prospect of a European future. Yet, Council Regulation 539/2001 of March 2001 blurred the euphoria over the integration in a broader European family, as it placed the new Balkan States (apart from Slovenia and Croatia) and Albania on a 'Black List' (list of countries whose citizens needed a visa to cross an external border of the EU).

The 2003 Thessaloniki Agenda announced the prospect of visa liberalization for the Western Balkans, and in this context cooperation in justice and home affairs became a priority. According to Schelter, the Western Balkan states, including Montenegro, were marked as a soft security threat for the EU (Schelter 2006). That is, the region was considered a belt for the transit of illegal immigrants, while some of its countries were related to organized crime. The conditionality related to the visa liberalization process was stipulated in the 2008 EU's Roadmap towards a Visa-Free Regime with Montenegro, which contained specific requirements (benchmarks) related to document security, tackling illegal immigration, re-admission, asylum policy, public order, judicial cooperation in criminal matters, and non-discrimination in ensuring its citizens the freedom of movement. The fulfilment of these benchmarks on the behalf of Montenegro eventually resulted in the visa-free regime for the country's citizens.

The second mechanism of EU's conditionality, which also showed the importance of the broader process of Europeanization, was related to the seven conditions that the EC recommended Montenegro fulfilled before opening accession talks. These conditions were enshrined in the Commission's Avis of November 2010, and were in line with the general observations of the state of the rule of law, democracy and human rights in Montenegro also noted in the reports of the Council of Europe and the OSCE (European Commission 2010). The first and the foremost requirement was for Montenegro to align its Constitution and its election legislation, which required a series of political compromises that the domestic political elites had been unable to reach since the country became independent in 2006. The importance of regulating franchise in a country of less than half a million voters, in fact, obstructed the compromise on electoral legislation between the government and the opposition.[1] Yet, it was precisely the power of Europeanization, pushed through the EU's conditionality stipulated in the Avis, that forced the political actors in Montenegro to agree on the changes to the Election Law. The compromise was necessary in the context of Montenegro's 2011 Progress Report, which subsequently recommended the opening of accession negotiations.

Conclusion

The history of Montenegro's relationship with the EU since the disintegration of Yugoslavia was not a linear one. It underwent dramatic changes, not only due to shifts in the domestic political context, marked by multiple disintegrations of states that Montenegro was party to and the bloody conflicts that consumed the neighbouring countries for much of the 1990s, but also due to changes in the EU's capabilities to act, and its general approach to the Western Balkans. Hence, the full understanding of this multilayered and transformative relationship assumed the overview of the four major stages in which Montenegro interacted with the EU.

In the period from 1991 to 1997, the relationship between the EC/EU and Montenegro was a strained one, and was developed under the aegis of the Twelve's interaction with the FRY, of which Montenegro was a constituent republic. In fact, the disintegration of Yugoslavia created numerous difficulties

1 Any changes to the election legislation require the support of a two-thirds majority in the Parliament of Montenegro. Given the current composition of the Parliament, the support of the opposition parties (most of which were members of the former unionist camp) for changing the election law was necessary. Thus, the opposition used its bargaining leverage to obtain concessions related to the citizenship of people from other republics of the former Yugoslavia, and to the name of the subject of language in elementary school as prerequisites for the political compromise on the change of the election legislation.

for the Twelve. The capabilities of the EC were seriously impaired at the outbreak of the conflict due to the disharmonious actions of the Member States, especially regarding the recognition of the post-Yugoslav states. In this setting, and given Montenegro's involvement in the conflict through the FRY, the relationship of the smaller republic of the third Yugoslavia with the EC/EU was strained due to the isolationist policies of Milošević's Yugoslavia.

The subsequent increased involvement of the EU in stabilizing the region has somewhat redeemed the failures of the CFSP in the early 1990s. This has been mirrored in the shift of the EU's policies towards Montenegro from 1997 to 2000 and in particular during the crisis in Kosovo in 1998 and 1999. Due to the departure of a portion of the Montenegrin elites from the regime in Belgrade, the relationship between the republic and the EU took on another note. Montenegro became the recipient of the EU's financial assistance, which was granted on the grounds of their support for the DPS's policy of countering Milošević and to aid the republic to take care of the people who sought refuge on Montenegrin soil at the time of the Kosovo crisis. Such aid, as noted in the respective section, provided an indirect push to the DPS's 'creeping independence' and the divisions within Montenegro over the continuation of the common state with Serbia.

It was precisely this division within Montenegro that marked the third period in the republic's relationship with the EU. After the departure of Milošević from power in Belgrade, the destiny of the common state became high on the Montenegrin agenda. At the same time, stability in the Balkan region and the projection of a single (or at least coherent) voice in international affairs after the Treaty of Nice were a priority for the EU. Hence, the conditionality principles of the EU played an important role at the time of the adoption of the Belgrade Agreement in 2002 and the negotiation of the referendum rules during 2005 and 2006. The desire of both camps in Montenegro to promote a pro-European agenda caused them to renegotiate their initial positions in order to meet the EU's requirements. At the level of European foreign policy, CFSP's involvement in the events related to Serbia and Montenegro increased the 'actorness' and 'presence' of the EU in external affairs. The enhanced coherence in the EU's foreign policy throughout this period has been paramount for the EU's redemption from its inactivity in the early 1990s, when the conflict in the former Yugoslavia erupted (Björkdahl 2005, Edwards 1997, Kintis 1993, Nuttall 1993).

The final stage in Montenegro's relationship with the EU is also the ongoing one, the one that was initiated after the country became independent in 2006. This period has been marked by an emphasis on the EU's conditionality and the overall process of Europeanization in Montenegro, which aspires to EU membership in the future. In the context of the EU, following the formalization of the conditionality of EU membership in the Thessaloniki Agenda of 2003, the EU became a norm exporter vis-à-vis the countries of the Western Balkans and thus Montenegro (Noutcheva 2007). As highlighted in the respective section, the direct and indirect impact of the EU's conditionality and Europeanization for the transformation and democratization of the Montenegrin political context were at their best during the visa liberalization process and at the stages when Montenegro sought to formalize its candidate status. The policy transfer in such cases included the adoption of a series of legislative changes which were previously hampered by the lack of domestic compromise. Hence, like the other aspiring members, Montenegro is in the process of acquiring and internalizing norms and values exported by the EU during its accession process. This means that Montenegro is subject to the process of Europeanization. In the context of the EU's overall relationship to the region, this implies that (at least for the aspiring members) the EU projects a single voice in the context of requirements for transition to democracy.

References

B92 2005. SCG: Terapija ili razvod. 23 February 2005. Available at: http://www.b92.net/info/vesti/index.php?yyyy=2005&mm=02&dd=23&nav_id=162872 [accessed: 13 September 2011].

B92 2006. Crna Gora kao talac. 3 May 2006. Available at: http://www.b92.net/info/vesti/index.php?yyyy=2006&mm=05&dd=03&nav_category=11&nav_id=196559 [accessed: 16 June 2011].

Baillot, M. 2004. Montenegro Eyes Parting from Serbia. *The Washington Times*, 21 September 2004.

Bieber, F. 2003. Montenegrin Politics since the Disintegration of Yugoslavia, in *Montenegro in Transition: Problems of Identity and Statehood*, edited by F. Bieber. Baden-Baden: Nomos Verlagsgesellschaft, 11–42.

Björkdahl, A. 2005. Norm-Maker and Norm-Taker: Exploring the Normative Influence of the EU in Macedonia. *European Foreign Affairs Review*, 10(2), 257–78.

Börzel, T.A. and Risse, T. 2000. When Europe Hits Home: Europeanization and Domestic Change. *European Integration online Papers*, 4(15). Available at: http://eiop.or.at/eiop/pdf/2000–015.pdf [accessed: 10 December 2011].

Buller, J. and Gamble, A. 2002. Conceptualising Europeanisations. *Public Policy and Administration*, 17(2), 4–24.

Dan 2005. EU: Crna Gora ne može opstati sama. 25 September 2005.

Djurić, D. 2004. Montenegro's Prospects for European Integration: On a Twin Track. *South-East Europe Review*, 4, 79–106.

Djurović, G. 2005. Pozitivna energija. *Euromont*, 31 May 2005.

Economist, The 2004. Semi-Independent. 10 April 2004.

Edwards, G. 1997. The Potential and Limits of the CFSP: The Yugoslav Example, in *Foreign Policy of the European Union: From EPC to CFSP and Beyond*, edited by E. Regelsberger, P. de Schoutheete de Tervarent and W. Wessels. London: Lynne Rienner, 173–95.

EU Council 2003. *Thessaloniki European Council, 19–20 June 2003: Presidency Conclusions*. Brussels, 1 October 2003. 11638/03.

European Commission 2005. *Communication from the Commission on the Preparedness of Serbia and Montenegro to Negotiatiate a Stabilisation and Association Agreemement with the European Union*. Brussels, 12 April 2005. COM(2005)476.

European Commission 2010. *Commission Opinion on Montenegro's Application for Membership of the European Union*. Brussels, 9 November 2010. COM(2010)670.

European Commission 2011. *EU Commission – Enlargement: Montenegro*. Available at: http://ec.europa.eu/enlargement/candidate-countries/montenegro/relation/index_en.htm [accessed: 10 November 2011].

European Council on Refugees and Exiles 2001. *Montenegro Country Report 2001*. Brussels: ECRE.

European Stability Initiative 1999. *Montenegro: A Balancing Act*. Brussels: ESI.

European Stability Initiative 2000. *Autonomy, Dependency, Security: The Montenegrin Dilemma: ESI Report*. Podgorica and Berlin: ESI.

European Stability Initiative 2001. *Sovereignty, Europe and the Future of Serbia and Montenegro: A Proposal for International Mediation*. Brussels: ESI.

European Stability Initiative 2005. *The Helsinki Moment: European Member-State Building in the Balkans*. Brussels: ESI.

Glas javnosti 2004. Radikali traže da se Crna Gora izjasni. 1 March 2004. Available at: http://arhiva.glas-javnosti.co.yu/arhiva/2004/03/01/srpski/vest-dana.shtml [accessed: 8 June 2011].

Goetz, K. 2000. Europeanising the National Executive? Western and Eastern Style. *UACES 30th Annual Conference*. Budapest, 6–8 April 2000.

Government of the Republic of Montenegro 2002. *Sporazum o principima odnosa Srbije i Crne Gore u okviru državne zajednice*. Available at: www.gov.me/files/1052297878.doc [accessed: 17 June 2011].

Grabbe, H. 2001. How Does Europeanisation Affect CEE Governance? Conditionality, diffusion and diversity. *Journal of European Public Policy*, 8(4), 1013–31.

Green Cowles, M., Caporaso, J. and Risse, T. 2001. *Transforming Europe: Europeanization and Domestic Change*. Ithaca, NY: Cornell University Press.

Gros, D., Kernohan, D., Najman, V. and Wozniak, P. 2010. On the Cost of Re-Establishing Yugoslavia: First Estimates of the Economic Impact on Montenegro of Establishing a Customs Union with Serbia. *Southeast European and Black Sea Studies*, 4(1), 119–39.

Hill, C. 1998. Closing the Capabilities-Expectations Gap, in *A Common Foreign Policy for Europe?*, edited by J. Peterson and H. Sjursen. London: Routledge, 18–38.

Hughes, J., Sasse, G. and Gordon, C. 2001. Enlargement and Regionalization: The Europeanization of Local and Regional Governance in CEE States, in *Interlocking Dimensions of European Integration*, edited by H. Wallace. Basingstoke: Palgrave Macmillan, 145–78.

Hughes, J., Sasse, G. and Gordon, C. 2004. *Europeanization and Regionalization in the EU's Enlargement to Central and Eastern Europe: The Myth of Conditionality*. Basingstoke: Palgrave Macmillan.

International Crisis Group 1999. Montenegro: Calm before the Storm. *Europe Briefing*, No. 10.

International Crisis Group 2000. Current Legal Status of the Federal Republic of Yugoslavia (FRY) and of Serbia and Montenegro. *Balkans Report*, No. 101.

International Crisis Group 2005. Montenegro's Independence Drive. *Europe Report*, No. 169.

Jovanović, I. 2005. Ex-Ex-Yugoslavia: A Virtual Country. *Transitions Online*, 7 March 2005.

Kintis, A. 1993. The EU's Foreign Policy and the War in Former Yugoslavia, in *Common Foreign and Security Policy*, edited by M. Holland. London: Pinter, 147–73.

Knill, C. and Lehmkuhl, D. 1999. How Europe Matters: Different Mechanisms of Europeanisations. *European Integration online Papers*, 3(7). Available at: http://eiop.or.at/eiop/texte/1999–007.htm [accessed: 10 March 2012].

Lippert, B., Umbach G. and Wessels, W. 2001. Europeanisations of CEE Executives: EU Membership Negotiations as a Shaping Power. *Journal of European Public Policy*, 8(6), 980–1012.

Lukić, R. 2003. The Painful Birth of the New State: Union of Serbia and Montenegro. *Conference: Southeastern Europe, Moving Forward.* Ottawa, 23–24 January 2003.

Lukić, R. 2005. From Yugoslavia to Serbia and Montenegro, in *Serbia since 1989*, edited by S.P. Ramet and V. Pavlaković. Seattle and London: University of Washington Press, 55–94.

Malešević, S. and Uzelac, G. 2007. A Nation-State without the Nation? The Trajectories of Nation-Formation in Montenegro. *Nations and Nationalism*, 13(4), 695–716.

Medojević, N. 2001. *Privatizacija i korupcija u Crnoj Gori*. Podgorica: PG Press.

Medjak, V. 2004. *Twin-Track Approach towards Serbia-Montenegro: A Recipe for Faster EU Integration*. Available at: http://www.sant.ox.ac.uk/ext/seesox/OpinionPieces_files/63.pdf [accessed: 22 July 2011].

Meurs, W. van 2003. The Belgrade Agreement: Robust Mediation between Serbia and Montenegro, in *Montenegro in Transition: Problems of Identity and Statehood*, edited by F. Bieber. Baden-Baden: Nomos Verlagsgesellschaft, 63–82.

Noutcheva, G. 2007. Fake, Partial and Imposed Compliance: The Limits of the EU's Normative Power in the Western Balkans. *Centre for European Policy Studies Working Document No. 274*. Brussels: CEPS.

Noutcheva, G. and Emerson, M. 2004. Europeanisations as a Gravity Model of Democratisation. *Centre for European Policy Studies Working Document No. 214*. Brussels: CEPS.

Nuttall, S. 1993. The EC and Yugoslavia: Deus ex Machina or Machina sine Deo?, in *The European Union 1993: Annual Review of Activities*, edited by N. Nugent. Oxford: Blackwell for the *Journal of Common Market Studies*, 11–25.

Obala Production 2004. *Rat za Mir.* Directed by Koča Pavlović.

Phillips, J. 2001. Montenegro Split by Ballot. *The Times*, 24 April 2001.

Pobjeda 2006. Predsjednik Skupštine Ranko razgovarao sa ambasadorima Francuske i Velike Britanije u SCG, Ig Perneom i Dejvisom Gaumanom: Konsenzus stvar političke volje. 31 January 2006.

Radaelli, C.M. 2000. Whither Europeanization? Concept Stretching and Substantive Change. *European Integration online Papers*, 4(8). Available at: http://eiop.or.at/eiop/texte/2000–008.htm/ [accessed: 24 August 2011].

Radonjić, R. 1998. *Tranzicije*. Podgorica: CID.

Rancharan, B.G. 1997. *The International Conference on the Former Yugoslavia: Official Papers*. The Hague: Kluwer Law International.

Roberts, E. 2002. *Serbia-Montenegro: A New Federation?* Zurich: Conflict Studies Research Centre.

Roche, D. 2007. Address at the opening of the 56th International Session of the European Youth Parliament. St. Patrick's Hall, Dublin Castle, 12 November 2007. Available at: http://www.reformtreaty.ie/speeches/rocheaddress.asp [accessed: 5 June 2011].

Schelter, K. 2006. Challenges for non- (and not yet-) Schengen Countries, in *Borders and Security Governance: Managing Borders in a Globalised World*, edited by M. Caparini and O. Marenin. Geneva: DCAF, 147–67.

Schimmelfennig, F. and Sedelmeier, U. 2005. The Europeanisations of Central and Eastern Europe. Ithaca, NY: Cornell University Press.

Šuković, D. 2002. Montenegrin Reactions to the Signing of the Agreement on the Union of Serbia and Montenegro. AIM Press, 18 March 2002. Available at: http://www.aimpress.ch/dyn/trae/archive/data/200203/20323–001-trae-pod.htm [accessed: 22 July 2011].

Teokarević, J. 2003. EU Accession and the Serbian-Montenegrin Constitutional Charter. *Romanian Journal of Political Science*, 3(2), 39–49.

Tocci, N. 2004. EU Intervention in Ethno-Political Conflicts: The Cases of Cyprus and Serbia-Montenegro. *European Foreign Affairs Review*, 9(4), 551–73.

Uzelac, S. 2003. Corruption in Transition Countries: 'How to Capture a State – The Example of Montenegro'. *South-East Europe Review*, 1(2), 103–15.

Vijesti 2003. Samostalno bi i brže i lakše u EU. 9 June 2003.

Vijesti 2006. Referendum i bez dogovora. 22 January 2006.

Wæver, O. 1998. Insecurity, Security and Asecurity in the Western- European Non-War Community, in *Security Communities*, edited by E. Adler and M. Barnett. Cambridge: Cambridge University Press, 69–118.

Chapter 6

From Deep-Seated Animosity to Principled Disagreement: A Comparative Analysis of Party-Based Euroscepticism in Serbia

Marko Stojić

Suspicion and scepticism towards Europe and the West in general have been a constant feature of a considerable part of Serbian society and politics over the last two decades. Such sentiments particularly flourished at the time of nationalistic euphoria and the wars that raged across the former Yugoslavia throughout the 1990s. While in other Central and Eastern European states 'returning to Europe' (Batory 2008b, Henderson 2008) was a key foreign policy objective and a common theme as a symbol of democracy and prosperity, the leading Serbian parties at that time – the Socialist Party of Serbia and the Serbian Radical Party – were Eurosceptic and nationalist, and their political agenda had nothing to do with potential Serbian accession into the EU. However, in the period after 2000, the political scene in Serbia has experienced a fundamental transformation of a type that is rarely seen in countries with a longer tradition of democratic institutions and multi-party political systems. Some of the hardest nationalists have become vocal advocates of Serbian membership of the EU, while at the same time some of the democratic parties that overthrew Milošević's regime have turned into key opponents of further Serbian EU accession.

This chapter examines the causes and consequences of the pronounced Eurosceptic sentiments among Serbian political elites. It aims to demonstrate how the question of European integration as a salient political issue has shaped a party system, as well as how it has been used by Eurosceptic parties in their policies and rhetoric since 2000. It also examines whether party ideology or strategy is a key driver of their positions on Europe. In order to capture peculiar features of Serbian party-based Euroscepticism, this analysis employs a framework that draws on and combines the two most widely used theoretical concepts, designed by Szczerbiak and Taggart (2008) and Kopecky and Mudde (2002). It examines party attitudes towards Europe, conceptualized as positions on the EU and positions on Serbian EU membership, and classifies parties into four distinct categories: Hard Eurosceptics and Hard Euroenthusiasts as well as Soft Eurosceptics and Soft Euroenthusiasts. The empirical focus of the chapter is on the Hard Eurosceptic Serbian Radical Party and the Soft Democratic Party of Serbia, and a group of

Soft Euroenthusiastic parties – the Serbian Progressive Party, the Socialist Party of Serbia and the New Serbia.

The study argues, firstly, that Serbian party-based Euroscepticism is a consequence of traditional and deep-seated animosity and suspicion towards Europe and the West in general (Serbian Radical Party) as well as principled disagreement with the policy of the EU and its key member states towards the former Yugoslavia, and in particular with regard to the issue of Kosovo's status (Democratic Party of Serbia). Secondly, parties characterized as 'hard' proponents or opponents of the EU or Serbian EU integration have been consistent over time and their positions are primarily ideologically driven. On the other side, parties that belong to the 'soft' categories have largely adopted (and changed) positions according to strategic and electoral considerations. The decision of these parties to shift stances on Serbian EU membership is thus a highly pragmatic one, aimed at maximizing chances of coming to power. Finally, party positions on EU membership tend to be much more easily shifted than their positions on the EU itself. The fact that it is harder for parties to change their attitudes towards the EU may indicate that such attitudes are more ideologically grounded and, as such, may reveal a real underlying position on Europe, unlike their stances on joining the EU, which have proved to have a more practical and strategic basis.

The analysis is based on parties' public statements, parliamentary voting on key European issues, as well as their programmatic documents. It also draws on interviews conducted with senior party officials from all relevant parliamentary parties. The chapter first briefly reviews the theoretical literature and then outlines a conceptual framework to be used in the analysis. The following section examines in greater detail the individual party attitudes and policies with regard to the EU. The final conceptual and empirical findings on the nature of party-based Euroscepticism in Serbia are summarized in the conclusion.

Conceptual Framework

The national parties' responses to Europe have recently become the topic of a growing body of literature. The literature addressing the issue of party-based Euroscepticism in Central and Eastern Europe has particularly flourished (Batory 2008a, Henderson 2008, Szczerbiak 2008), although a large majority of authors deal with individual case studies and rarely offer a conceptual framework for analysis of Eurosceptic party stances. Nevertheless, two theoretical approaches to party attitudes to Europe stand out as the most widely used in the comparative literature, namely the concept of 'hard' and 'soft' Euroscepticism proposed by Szczerbiak and Taggart (2008), and a categorization of both positive and negative party positions on Europe designed by Kopecky and Mudde (2002).

Szczerbiak and Taggart broke down negative party positions on Europe into Hard and Soft Euroscepticism. The first term initially implied 'a principled opposition to the EU and European integration and therefore can be seen in parties who think that their countries should withdraw from membership, or whose policies towards the EU are tantamount to being opposed to the whole project of European integration as it is currently conceived' (Szczerbiak and Taggart 2008: 2). The second terms referred to a party position 'where there is not a principled objection to the European integration or EU membership, but where concerns on one (or a number) of policy areas lead to the expression of qualified oppositions to the EU, or where there is a sense that "national interest" is currently at odds with the EU trajectory' (Szczerbiak and Taggart 2008: 2). These authors later reformulated this concept and abandoned the idea that attitudes towards a country's membership of the EU should be viewed as the ultimate litmus test of whether a party belongs to the Hard or Soft Eurosceptic camp. They specifically argued that the key variables in determining party attitudes should be underlying support for or opposition to the European integration project as embodied in the EU, as well as attitudes towards further actual or planned extensions of EU competencies, rather than a party's stances to their country's membership at any given time (Szczerbiak and Taggart 2008: 242).

On the other hand, the concept developed by Kopecky and Mudde (2002) covered both pro- and anti-EU positions and introduced a distinction between support for the ideas of European integration that underlie the EU and support for the EU as it is in reality. They further made a distinction between Europhiles and Europhobes based on support for, or opposition to, the ideas of European integration, as well as between EU optimists and EU pessimists based on the attitudes to the EU as it is or as it is developing. As a consequence, the authors constructed the four ideal-type categories of party positions on Europe. Euroenthusiasts thus combine Europhile and EU-optimist positions and support both the general ideas of European integration and the EU as it is. Eurosceptics express Europhile and EU-pessimist positions and support the general ideas of European integration but are pessimistic about the EU as such. Eurorejects adopt Europhobe and EU-pessimist attitudes and oppose both the ideas underlying the process of European integration and the EU. Finally, Europragmatists combine Europhobe and EU-optimist positions as they do not support the general ideas underlying the EU but support the EU.

These theoretical frameworks have been successfully applied by their authors in the case of Central and Eastern European (CEE) countries. Moreover, a number of other scholars have used the same frameworks, particularly the concept of Soft and Hard Euroscepticism. These authors often adapted the original concepts in order to fully capture specific features of individual case studies. Henderson (2008) thus analysed Slovak parties by adding a category of 'phony Europhiles' to capture those parties that are strategically in favour of EU integration, but reluctant to face that this has implications for domestic politics and that it requires an engagement with the realities of Western democratic principles. Conti (2003) analysed Italian parties and introduced a neutral category, defined as lacking a clear position on European integration, which is the position of a party in transition between past and new positions or of a recently born party on the way to shaping its preferences.

This study similarly argues that these two key concepts may also be employed in the Serbian case, but that they should be slightly adapted in order to grasp all the peculiar features of Serbian party stances, which are the result of the fact that the EU is perceived differently in a candidate state with rather specific relations with the EU. It therefore applies the original concept of Hard and Soft Euroscepticism, which, unlike the reformulated concept, takes into consideration party stances on EU membership and consequently, in the context of Serbian party politics, more accurately classifies their positions. Hard Eurosceptic parties in the case of Serbia thus argue that the country should not join the EU (withdraw from membership in the original concept) and oppose, although less vocally, the whole project of European integration (Serbian Radical Party). Soft Eurosceptic parties, while they do not express principled objection to European integration or EU membership, argue that the 'national interest', that is preservation of Kosovo within Serbia, is at odds with Serbian EU membership (Democratic Party of Serbia).

On the other side, the concept of Kopecky and Mudde (2002) requires rather nuanced and clearly differentiated party positions on ideas of European integration that underlie the EU as well as the EU as it is in reality. However, Serbian parties have not publicly elaborated such attitudes (neither in party rhetoric nor programmatic documents), although when directly asked about it, the interviewed party officials did express strong opinions. Szczerbiak (2008: 239) also pointed to this issue when analysing Polish parties in the pre-accession period and argued that 'there was simply not enough data available at that time, either from the party's published policy statements or from statements by its representative leaders, to be able to categorize the party as being Eurosceptic or not'. This can be explained by the fact that citizens and voters know almost nothing about the EU, and political parties have no incentive to exploit these issues, unlike the issue of Serbia's membership in the EU, on which almost every citizen has an opinion. As a result, all Serbian parties have adopted elaborated positions on Serbian membership in the EU.

In order to allow a comparative analysis with parties in other CEE states, but at the same time to present the peculiar features of party stances on Europe in the context of Serbian party politics, the study combines these two approaches. The rationale behind this lies in that, while the first approach, when adjusted to a candidate state, may account for Hard and Soft Eurosceptic parties, it may not explain the position of former Eurosceptic parties and why some parties changed attitudes towards Europe. Therefore, the study also adapts and employs Kopecky and Mudde's (2002) concept. It specifically proposes the similar four-fold categorization of parties, but it reformulates key indicators on which the classification of parties has been based. First, given that Serbian parties have not had clearly defined attitudes towards the ideas that permeate European integration or the EU as it is, these two indicators are grouped into one category labelled 'the attitude towards the EU', to encompass all possible and not fully differentiated positions on the EU. On the other hand, the concept uses 'the attitudes towards

Serbian EU membership', which is a priority for all parties in the country, as another important indicator of their overall attitudes towards Europe.

Drawing on these two theoretical models, the concept combines negative and positive attitudes towards the EU and Serbian EU membership, and groups Serbian parties into four basic categories that may account for their complex attitudes towards Europe (Table 6.1). The group of parties that supports Serbian EU membership and the EU is therefore characterized as Hard Euroenthusiasts, while parties that expressed a negative position on both issues are termed Hard Eurosceptics. On the other side, parties that support Serbian EU membership, but have a neutral position, or combine both positive and negative stances, or have critical although not opposing attitudes to the EU are termed Soft Euroenthusiasts, while those with similar attitudes to the EU, but opposed to Serbian EU membership are characterized as Soft Eurosceptics.

Table 6.1 Party positions on Europe

		Position on the EU		
		Positive	**Positive/No opinion/ Negative**	**Negative**
Position on EU membership	**Positive**	Hard Euroenthusiasm	Soft Euroenthusiasm	
	Negative		Soft Euroscepticism	Hard Euroscepticism

Source: Kopecky and Mudde 2002, Szczerbiak and Taggart 2008.

This concept also aims to address another key issue discussed in the comparative literature on Euroscepticism – whether parties' positions on Europe are ideologically- or strategically-driven. The study argues that both factors may have explanatory value in the Serbian case and identifies two key tendencies. First, it appears that positions of parties categorized as Hard Eurosceptic and Hard Euroenthusiasts are primarily ideologically driven. These parties have not markedly changed their positive (or negative) position on Europe since they were founded, despite sometimes moderated rhetoric employed by some Hard Eurosceptic parties for strategic, electoral reasons. The deeper rootedness of their positions may be also seen in the fact that there is no contradiction in their positions, since they have consistently supported (or opposed) both the EU and Serbia's EU membership. Hard Euroscepticism perceived in this way is in line with the original

Szczerbiak and Taggart's (2008) concept of principled opposition to the EU and the country's EU membership, while it is similar to the Euroreject category of Kopecky and Mudde (2002), although not the same, since these authors disregard EU membership as an indicator of party positions on Europe.

On the other hand, the attitudes of parties termed Soft Eurosceptic and Soft Euroenthusiastic have primarily been pragmatically formed. These parties are mainly guided by strategic, electoral considerations rather than ideological principles. This has been evidenced by the fact that they have fundamentally changed positions on Serbian EU membership in the past decade in an attempt to gain political power. Soft Euroenthusiasts thus, as noted by Kopecky and Mudde (2002: 303) with regard to the Europragmatist category, 'do not hold a firm ideological opinion on European integration, and decide to assess EU (membership) positively on the basis of pragmatic considerations', because they deem it profitable for their country or electorally desirable. In addition, their stances on the EU are not elaborated for strategic reasons to mask often negative stances in order to legitimize themselves as pro-European parties. Therefore, their position on the EU is a combination of neutral and positive (Socialist Party of Serbia), and neutral and negative (Serbian Progressive Party, New Serbia) attitudes. Soft Euroenthusiasm is also devoid of the illogical feature of the Europragmatist category, that is that parties opposed to EU integration in principle are supportive of the EU's current trajectory (Szczerbiak and Taggart 2008), because it is possible and politically viable that a party strategically supports EU membership despite opposition to, or a lack of attitude towards, the EU. Given the strategic nature of their attitudes and a lack of grounded stances on the EU, these parties may become Soft Eurosceptic and start advocating opposition to a country's EU membership. Such parties do not express principled objection to the EU or EU membership, but decide to focus on the prevailing Eurosceptic electorate, exploit their concerns for the country's EU membership deal and express qualified opposition to the country's EU accession, in order to secure more votes (New Serbia between 2008 and 2010). In addition, the position of the Democratic Party of Serbia, elaborated later in the study, is characterized as Soft Euroscepticism, although it is rather unique, since its opposition to Serbian EU membership is not strategically driven.

Individual Party Positions

The next section deals with individual party positions on Europe (Table 6.2). As conceptualized above, party attitudes are examined in relation to both the EU and Serbian EU membership. The attitudes of Hard and Soft Eurosceptic parties are first analysed, followed by the study of parties that have shifted stances on Europe in 2008 and are thus termed Soft Euroenthusiasts.

Table 6.2 Serbian party positions on Europe (April 2012)

		Position on the EU		
		Positive	**Positive/No opinion/Negative**	**Negative**
Position on EU membership	**Positive**	*Hard Euroenthusiasm* Democratic Party G17 plus Liberal Democratic Party	*Soft Euroenthusiasm* Serbian Progressive Party Socialist Party of Serbia New Serbia *Democratic Party of Serbia until 2008*	
	Negative		*Soft Euroscepticism* Democratic Party of Serbia *Socialist Party of Serbia until mid-2000s* *New Serbia from 2008 to 2010*	*Hard Euroscepticism* Serbian Radical Party

Eurosceptic Political Parties

Serbian Radical Party: Deep-seated animosity towards the EU

The Serbian Radical Party (SRS) is a radical right and nationalistic party whose fundamental political aim was 'the unification of the entire Serbian nation and establishment of a single, unitary state, a Great Serbia, on the whole Serbian national territory, which would include Serbia, Montenegro, the Republic of Srpska and the Republic of Serbian Krajina' (SRS 2009: 2). Its programme (SRS 2009) also advocated development of national consciousness and patriotism, preservation of national traditions, protection of the traditional Serbian family and bringing up the youth in Serbian Orthodox spirit. The party has always expressed firm opposition to Serbian integration into the EU, despite occasional moderation of its discourse in the mid-2000s. While its position on the EU was less pronounced in public discourse and programmatic documents, it was also strongly negative. Therefore, the Serbian Radical Party may be termed a Hard Eurosceptic party (Table 6.2), given its principled and ideologically driven objection to both the EU and Serbian EU membership.

Attitudes toward the EU were largely absent from the party documents and public rhetoric. The party did not express in its programmatic documents fully elaborated attitudes towards the EU, although its programme demonstrated the party's anti-globalist character by pledging 'to make Serbia a member of international organizations that respect its interests and the principle of non-interference in internal affairs of sovereign countries' (SRS 2009: 29). The party thus proclaimed that it was in favour of international organizations 'whose activities are based on cooperation and solidarity, rather than pressure, blackmail and force' (SRS 2009: 29). The absence of specifically anti-EU discourse can be explained by the fact that the essence of European integration was unknown to the Serbian public and voters, and therefore criticism of the EU could not bring considerable political gains, unlike a strong party objection to the EU's policy towards Serbia.

However, following the 2008 financial crisis, one can observe this party's attempt to present the EU as a completely failed experiment of the western European elites that exclusively serves the economic interests of the rich Member States. Martinović, the party MP and a deputy president of the parliamentary group (Velika Srbija 2011: 7), argued that 'the EU is a political, economic and moral corpse, given that Greece is technically bankrupt, and that a similar scenario is almost inevitable in Ireland, Italy and Spain'. The Radical MP Mirović (2011) similarly pointed out that 'the EU is only useful for Western European members, while it is absolutely useless for Eastern Europe. The very existence of the EU is the result of the interests of the largest exporters and in any case it is not in the interest of small nations'.

The Radicals support 'the De Gaulle principle of a Europe of nations from the Urals to the Atlantic', since they argue 'Europe cannot be whole without Russia' (Martinović 2011), while their preferred form of European organization is Europol and the Council of Europe, because 'the Russians are also there' (Mirović 2011). Interestingly, they also argue that the European Parliament is a second-rate institution, while the European Council is an undemocratic body given that 'there is not a similar system in the world, where the laws are passed by the ministers' (Mirović 2011). Martinović (2011) concluded that 'the EU is too cumbersome and bureaucratic organisation that has lost the purpose of its existence'.

On the other side, the Serbian Radical Party has always been a principled opponent of Serbian accession to the EU. Such a position was primarily ideologically driven and deeply rooted, and as such has been consistent over the time. The party, however, experienced a moderation of its public rhetoric in the mid-2000s, but it has never fundamentally changed policies. The case of this party demonstrates the analytically significant difference, identified by Szczerbiak and Taggart (2008), between a party's underlying position on the EU (and EU membership in this case) and its public rhetoric, which may be the result of strategic considerations and not necessarily an indicator of the real party position.

This party position on Serbian EU membership was directly related to intra-party power struggles. When in 2003, a party leader, Vojislav Šešelj, went to The Hague to stand trial for war crimes at the International Criminal Tribunal for the former Yugoslavia (ICTY), a party deputy president, Tomislav Nikolić, and a secretary-general, Aleksandar Vučić, took over the party leadership. The party managed to win the 2003, 2007 and 2008 elections and become individually the largest party in the parliament, securing up to 29 per cent of the total votes (Table 6.3). The electoral success of the Radicals was primarily due to a moderated nationalistic rhetoric that emphasized socioeconomic issues, such as unemployment, privatization and corruption, rather than nationalism. In addition, the Radicals strategically opted to tone down its opposition to Serbian EU membership, given that a large majority of voters supported Serbian EU accession. Nikolić even declared in the 2003 presidential election campaign that 'he and his party will provide a full contribution to Serbian accession to all European institutions and organizations, particularly the EU, but by preserving Serbian identity, national pride, honour and dignity' (Komšić 2007: 15). The case of the Radicals in the mid-2000s therefore demonstrates the strong electoral incentive to moderate pronounced Eurosceptic rhetoric even when a party is deeply ideologically grounded in anti-Europeanism.

Table 6.3 Electoral results of (former) Eurosceptic parties

Political party	**Number of MPs and % of votes**				
	Dec 2000	*Dec 2003*	*Jan 2007*	*May 2008*	*May 2012*
	Eurosceptic parties				
Serbian Radical Party	23 9%	82 28%	81 28%	78 29% (57, April 2012)[1]	0 4.6%
Democratic Party of Serbia	–[2]	53 18%	47 16%[3]	30 11%[4]	21 6.8%
	Former Eurosceptic parties				
Serbian Progressive Party	–	–	–	– (21, April 2012)	73 24%[5]
Socialist Party of Serbia	37 14%	22 8%	16 5%	20 7%[6]	44 14.7%[7]
New Serbia	–	22 8%[8]	–	–	–

Source: Cesid 2012, Emerson and Šedo 2010: 14, National Assembly of the Republic of Serbia 2012.

1 Twenty-two MPs left the Serbian Radical Party in September 2008 and formed the Serbian Progressive Party. One of the breakaway MPs returned to the Serbian Radical Party in February 2011.

2 The Democratic Party of Serbia and New Serbia were part of the pro-European Democratic Opposition of Serbia.

3 In coalition with New Serbia.

4 In coalition with New Serbia.

5 In coalition with New Serbia.

6 In coalition with United Serbia and the Party of United Pensioners of Serbia.

7 In coalition with United Serbia and the Party of United Pensioners of Serbia.

8 In coalition with the Serbian Renewal Movement.

The strategic nature of the Radicals' rhetoric shift was, however, visible at times when electoral concerns were less important, that is in the period immediately after the elections. For instance, in May 2007, Nikolić, as a temporary speaker of the parliament, argued that Serbia should seek closer ties with Russia and not the West. He said that Russia would bring together 'nations that will stand up against the hegemony of America and the European Union' and that 'Serbia should associate itself with the Russian and Belarusian union' (*RFERL* 2007a). He also claimed that a majority in Serbia would strive for membership in a Russian-led alliance of states and not in the EU and added that 'unfortunately Serbia was not a Russian province' (*RFERL* 2007a). The shift in rhetoric was recognized by the EU, whose enlargement commissioner Olli Rehn warned that 'the election of an ultra-nationalist as Serbia's parliamentary speaker is a worrying sign' and that Serbia was 'at a crossroads' (*RFERL* 2007b). Finally, in the words of Mirović (2011), at that time the party had 'a rather populist and pragmatic rhetoric', arguing that Serbia could join the EU, provided that the EU recognized that Kosovo was part of Serbia, 'due to popularity of the EU with the Serbian electorate'. He added that this was the pure pragmatism of a former party leadership, based on the results of public opinion research, which showed that a majority of citizens were in favour of the EU, and not a result of the belief that it was in the interest of Serbia.

Following one more failed attempt to form a government in 2008, and faced with the fact that moderated rhetoric did not bring tangible results (mainly due to the party's a lack of coalition potential), Nikolić became aware that if the party was to take political power, it must fully reorient its policies. However, that was not possible under the leadership of charismatic party leader Šešelj, who had overwhelming support for his nationalistic and anti-European policy. This consequently led to a party split and the foundation of the new Serbian Progressive Party. After a major intra-party schism in September 2008, which shook the very foundations of the party, the Radicals reaffirmed their principled position – an absolute and unconditional opposition to Serbian EU integration.

The Serbian Radical Party's opposition to Serbian integration into the EU was primarily ideologically driven, given that the party was deeply grounded in the pronounced ideology of anti-Westernism. The party programmatic documents, as well as the discourse of its leader Šešelj, demonstrated traditional scepticism, a deep-seated animosity, and even hatred towards the EU and the West in general. In his 2006 political testament, Šešelj called on party members to 'strongly oppose any attempts to include Serbia in NATO and the EU, because all traditional Serbian enemies are there' (Komšić 2007: 14). He went on to argue that 'they (members of the EU) have been furious with us because the Serbs had defeated their grandfathers and ancestors who therefore left a testament to their heirs to punish Serbia' (Komšić 2007: 16).

The Radicals argued that EU candidacy status was a poor reward 'considering how many of our children, mothers and fathers, and brothers and sisters were killed by criminals from NATO and leading members of the EU, and when we remember how many of our heroes were extradited to the Hague dungeon' (Velika

Srbija 2011: 8). The party believed that the EU, through its rule-of-law mission in Kosovo, directly created an independent state on the territory of Serbia. It called on the government to suspend all negotiations with the EU and 'to give up the disastrous policies of European integration and turned to the countries that respect international law and territorial integrity of Serbia' (SRS 2010). The party claimed that Serbian EU accession was 'a suicidal policy of destruction of the Serbian state' (Velika Srbija 2011: 7). The Radicals also accused the ruling pro-European coalition of having obedient and treacherous policies that led to territorial dismemberment, the giving up of Kosovo, and sacrificing the lives of Serbian people by orders of Brussels and Washington (Velika Srbija 2011: 7).

The Serbian Radical Party, in contrast, expressed a pronounced pro-Russian orientation. The party believed that gradual opening of the market for EU products, which is provided for in the Stabilisation and Association Agreement, would have extremely negative consequence for the economy and that 'the only salvation for Serbian foundering economy is in close connections with Russia' (Mirović 2011). It is also the only relevant party that advocated full Serbian membership of the Collective Security Treaty Organization, a military alliance of a number of former Soviet republics. In summary, the party politics was a reflection of traditionally anti-European sentiments, which have been a constant feature of a considerable part of Serbian society and politics over the last two decades. Moreover, the controversial policy of key EU members regarding the issue of Kosovo and related preconditions for Serbian EU accession further reaffirmed its principled anti-EU attitudes.

The study finally argues that tactical reasons may not be considered as a factor that impact on this party's attitude. Hard Euroscepticism was, however, an electorally profitable position since the 2008 proclamation of Kosovo's independence, which was widely perceived in the Serbian public as directly sponsored by key EU Member States. Serbia's EU integration was supported by 51 per cent of citizens in December 2011, which was the lowest support since 2002 (SEIO 2012), and the party therefore had pragmatic and strategic incentives to maintain, rather than change its Hard Euroscepticism. Nevertheless, the downward trend in support for Serbian EU accession may be seen as reinforcement, rather than a cause of party attitudes, given that it had essentially maintained such a policy even when the EU was more popular with the Serbian voters.

Democratic Party of Serbia: Principled disagreement with the EU

The Democratic Party of Serbia (DSS) is a conservative, centre-right party whose key principles include support to the Serbian Orthodox Church, preservation of traditional moral values, protection of national identity, strengthening of national cultural institutions and protection of the Cyrillic script (DSS 2010a). This party expressed a complex attitude towards the EU and Serbian EU membership. On the one hand, it demonstrated mistrust of the West, primarily the United States of America, and particularly contested the legitimacy of the ICTY (Goati 2009). On the other hand, the party supported Serbian EU accession and significantly contributed to this process while it was in power. However, recognition of Kosovo's independence by a large majority of EU Member States fundamentally affected party attitudes, and since 2008 it has argued for stopping further integration into the EU. The party position may be termed Soft Eurosceptic (Table 6.2), since it has never objected to the EU, while as a consequence of principled disagreement with the position of key EU Member States on the issue of the status of Kosovo, it strongly objected to Serbian EU membership.

This party did reflect on the EU, although that has not been pronounced in party politics and programmatic documents. The Democratic Party of Serbia perceived the EU's policy towards the Western Balkans in a broad context and as an indicator of the shortcomings of the EU's overall foreign and security policy. Specifically, the party vice president Samardžić (2011) argued that the EU does not have its own authentic security and foreign policy, and pointed out that it has become a periphery under the influence of the United States of America. Moreover, the party objected to the EU's policy towards the region, since 'it does not deal anymore with strengthening institutions, democratization and economic stabilization, but is rather a result of geo-strategic concerns that are directly related to NATO's intention to include all regional countries in its membership' (Samardžić 2011). The party thus expressed criticism but, unlike the Serbian Radical Party, did not object to the whole project of European integration, nor did it advocate an alternative form of cooperation among European nations.

On the other side, the Democratic Party of Serbia had a complex attitude towards Serbian integration into the EU. The party did not object to Serbian EU membership before 2008, and while it was a key ruling party in the period from 2004 to 2008, the country made some initial steps towards EU membership. The party president and then Serbian Prime Minister Koštunica stated in March 2004 that Serbia essentially belonged to Europe and that 'EU membership is not only what we want, but also something that has to be done and cannot be avoided', adding that there was no alternative to the European path (Government of the Republic of Serbia 2010). In May 2007, Koštunica's second government announced that its main short-term goal would be acquiring EU candidate status and speeding up the process of legislative harmonization with the EU standards (Parliamentary Network 2010).

The Democratic Party of Serbia opposed cooperation with the ICTY, which was a key precondition for EU accession, and favoured voluntary surrendering of the indicted. The government thus suspended cooperation with the ICTY, which resulted in withdrawal of financial support from the Western countries and postponement of the feasibility study on Serbia's readiness to enter into a contractual relationship with the EU. As a consequence of the negative economic effects of these policies and bad results at the presidential and local elections in 2004, the party gradually modified its attitude toward the ICTY. The government thus managed to 'persuade' some of the indicted to surrender voluntarily and adopted the National Strategy for Accession to the EU as a sign of its readiness to continue European integration. This led to the EU's decision in mid-2005 to resume negotiations with Serbia. The negotiations were successfully completed and the Stabilisation and Association Agreement was initialled in October 2007, supported by the Democratic Party of Serbia.

However, after leading EU Member States recognized the independence of Kosovo in 2008, the party changed its position and adopted a new policy of opposition to Serbian EU integration, given that 'the EU violated the fundamental norms of international law' (Samardžić 2011). The party argued that under the radically new circumstances Serbia must refuse to sign the Stabilisation and Association Agreement, claiming that to sign it would implicitly recognize the independence of its southern province (Stojić 2010). After losing the May 2008 elections, it placed the question of Serbian EU integration in the absolute centre of its politics, and became the main critic of the government's pro-European policy as well as a strong opponent of Serbian European integration.

The party adopted the motto 'Serbia Has No Alternative!' and argued that 'instead of disastrous politics that the EU has no alternative, it is time for the state-building politics, whose fundamental principle is that Serbia, with Kosovo as its integral part, could join the EU' (DSS 2010b). Koštunica argued that 'there is not a single European state that has given up part of its territory under pressure or has recognised a violent change of its borders. The rules that apply to all European countries must also apply to Serbia' (DSS 2010b). In the run-up to the 2012 elections, the party further strengthened its Eurosceptic position and adopted a policy of military and political neutrality as an alternative to EU membership.

The Democratic Party of Serbia belongs to the Soft Eurosceptic category. This party had key characteristics identified by Szczerbiak and Taggart's (2008) initial concept of Soft Euroscepticism. It did not express principled objection to the EU or EU membership, while it had concerns on one of the EU's policy areas, namely the EU policy towards the Western Balkans, which de facto (although still not officially) required Serbia to accept Kosovo independence in order to become an EU member state. It therefore argued that the national interest of preservation of Serbian territorial integrity is at odds with EU membership, and as a consequence it was firmly against Serbian EU membership.

In addition, it is a Soft Eurosceptic party that fulfils the criteria conceptualized earlier, that is, it objected to Serbian EU integration, while not expressing completely negative attitudes towards the EU. This party position is, however, specific and unique in the way that it is not pragmatically and strategically driven (which may be a key characteristic of parties belonging to this category), and is rather a consequence of a deep traditional and national ideological conviction that territorial integrity and sovereignty is a prime European value. The Democratic Party of Serbia is not essentially a party with anti-European ideology. Its opposition to Serbian EU accession, although practically similar, is substantially different from the stance of the Serbian Radical Party. The party officials underlined its European orientation based on old European, conservative heritage and values, such as commitment to the rule of law and the market economy, and added that 'the party does not have any backup policy' (Samardžić 2011). Its programme also stated that Serbia should have close cooperation with European countries, with the ultimate goal of becoming an EU member state (DSS 2012: 11). The party advocated 'a new national policy that will have as main objectives Serbia itself and its internal development, but based on the best European values and standards that are in the interest of the country' (DSS 2010c). It also insisted on an open dialogue with the EU in order to build 'beneficial relations based on the mutually accepted agreement' (DSS 2010c).

Finally, it appears that strategic reasons do not play an important role in this party's position on Serbian EU integration, given that the Democratic Party of Serbia is widely perceived as 'a party that is not guided by political pragmatism' (Samardžić 2011). It is rather a value-based party that gives priority to its programmatic principles and national politics, even when faced with declining support in the Serbian electorate. Popularity and electoral success are not the most important goals to the party leader Koštunica, who is 'persistent and stubborn' when implementing principled politics (Bakić 2011). Bakić (2011) concluded that the party 'insufficiently rationally perceived the reality' and as result it faced a sharp decline in electoral support.

Former Eurosceptic Political Parties

Serbian Progressive Party: Unproven 'Europeans'

The Serbian Progressive Party (SNS) is a centre-right party, which was founded by a group of former moderate leaders of the Serbian Radical Party. The new party abandoned the Radicals' ideology of Serbian nationalism and the concept of a Great Serbia (SNS 2010a) and underwent a fundamental and rapid transformation with regard to attitudes towards Serbian EU membership. Even though the party leaders had expressed pronounced Euroscepticism for almost twenty years (with occasionally moderated rhetoric in the mid-2000s), they founded the Serbian Progressive Party on a radically new, pro-European platform and became firm advocates of Serbian EU membership. The party is characterized as a Soft Euroenthusiast (Table 6.2), due to its strategically driven support for Serbian EU integration. Its position on the EU cannot be definitely assessed, since it has never been elaborated. However, based on limited available data, it appears that this party is rather critical towards the EU, but strategically chooses not to speak about it.

The Serbian Progressive Party has never articulated attitudes towards the EU, since its focus was on Serbia's EU membership. One cannot expect any deeper reflection on the principles of European integration given that it is a new, pragmatic party, whose politics towards the EU is devoid of principled positions on the desirable form of a community that brings together European nations, which is the case with most political parties in Serbia. However, the fact that this party's key partner from the EU is Austria's radical right and populist Freedom Party (Fallend 2008) may indicate its position on the EU. Their cooperation agreement specifically stated that the two parties support 'the creation of a Europe of free nations and self-determined people in the framework of a grouping of national sovereign states' (SNS-FPÖ 2011). They required the conservation of national identities, including the Western traditions of Christianity, humanism and the Enlightenment, as well as 'effective protection of Europe against the guardianship of imperialist superpower'. The parties also stood for a fight against globalization and 'the infiltration of religious fanaticism into a European society'. It may therefore be argued that the Serbian Progressive Party shared the rather sceptical and negative stance of its partner on the EU. This, however, has never been stated in the party's public rhetoric or documents, since it strives to move away from its anti-European political legacy, present itself as a modern, conservative, pro-European party, and obtain 'European legitimacy'.

On the other side, the party leadership underwent a fundamental transformation regarding its attitudes towards Serbian EU membership. After leaving the Radicals in 2008, the party leader Tomislav Nikolić abandoned nationalistic and anti-European politics and finalized a transformation that he had rhetorically started as deputy president of the Serbian Radical Party. The new party immediately started advocating Serbian EU membership. Its programme stated that the party stood for Serbia's membership of the EU, military neutrality, intensified cooperation with the Russian Federation, China and Japan, as well as the best possible relations with the United States of America (SNS 2011b). It also argued that the party supported the European integration process aimed at institutional and economic strengthening of Serbia, and that it believed that Serbian EU accession was in the best, long-term interests of all citizens (SNS 2011b: 41). Its initial programmatic principles stated that Serbia could join Europe only as a whole state, with Kosovo as its integral part (SNS 2010a: 1). However, the reference to Kosovo in relation to the EU accession was later excluded from the party programme, which indicates that the party put priority on Serbian EU integration. Moreover, the party claimed that recognition of Kosovo's independence was not a condition for Serbian membership of the EU, and argued that if the EU ever conditioned Serbian EU accession in such a way, then the Serbian citizens should make a decision on this issue at a referendum (SNS 2010b).

The decision of the leaders of the Serbian Progressive Party to start advocating Serbian accession to the EU was highly pragmatic. It rested upon the fact that a vast majority of voters were in favour of Serbian EU membership, and that the party had no chance of winning the election if it did not follow a European path. It was therefore necessary to break away from the Serbian Radical Party, given that the Radicals had failed to get into power advocating anti-European politics in the period after 2000. The party's adoption of a radically new attitude towards the EU may thus be primarily interpreted as an election strategy and tactics to come to power, a fact that Nikolić himself did not hide. Thus, he argued that 'the Radicals have never had a desire to come to power, while the Serbian Progressive Party is something else. We are a pro-European party. If we stand against the EU, we will never be able to win the elections in Serbia' (SNS 2011a).

The party did not hold a firm ideological opinion on Serbian EU membership, and assessed it positively based on pragmatic considerations, which makes its Euroenthusiasm soft. It had a primarily utilitarian perception of Serbia EU membership, and viewed the EU as a main economic partner that could contribute to a better life for ordinary people. Its vice president Aleksandar Vučić (SNS 2010c) thus said about the EU: 'I do not care for them, I just respect them. I neither love them nor they are especially dear to me, but we, as responsible people, have to take care of our country'. He also added that 'we need a rational, realistic and serious approach to national politics in order to get the most we could for the country and to lose the least' (SNS 2010c). It appears that the party's Soft Euroenthusiasm was not deeply grounded in any principled beliefs in the intrinsic good of EU membership.

In addition, the Serbian Progressive Party was not fully ideologically positioned and consequently it could not have ideologically driven attitudes to the EU. It published its party programme more than three years after it was founded and it was reluctant to take a clear stance on a number of key and politically very delicate issues, such as the status of northern Kosovo as a condition for Serbian EU candidacy status. Many thus concluded that 'it is clear that they want to come to power, but no one could guess how they would behave and what kind of policy they would implement' (Cvijanović 2011, Djurković 2011: 208). What is certain is that the case of Serbian Progressive Party demonstrated the powerful transformative power of strategic electoral incentives that may even lead to a breakaway of parties and a fundamental change of their positions.

Socialist Party of Serbia: Emerging 'Europeans'

The Socialist Party of Serbia (SPS) was a ruling party throughout the 1990s, whose politics led to the international isolation of the country and NATO military intervention in 1999. This communist-successor party may be categorized at that time as the left-centre national populist party (Siaroff 2000), since it had been positioned left of the socioeconomic centre, but strong populist nationalism, rather than focus on socioeconomic issues, was its most important characteristic. Although anti-European and nationalist, the party did not articulate outright rejection of Serbian EU membership, given that at the time it was an important political issue, while it criticized the Western policy towards Serbia. As such, its position at that time is characterized as Soft Euroscepticism with some Hard Eurosceptic elements, although not as pronounced as in the case of the Serbian Radical Party. After 2000, the party embarked on a slow transformation in an effort to rebrand itself as a modern, pro- European social-democratic party. As a consequence, the Socialist Party of Serbia is now characterized as a Soft Euroenthusiastic party (Table 6.2) because of its pragmatically driven support for Serbian EU membership, and the lack of a publicly expressed position on the EU.

Until the mid-2000s, this party was largely perceived as a nationalist, anti-globalist and anti-Western party that was expressing a strong critical stance on the EU, and particularly the United States of America. In 2002, the party 'condemned

Europe because of its participation in the 1999 aggression against Yugoslavia, which was an expression of American imperialist strategy' (Komšić 2007: 28). The Socialists argued that 'Europe has participated in the destruction of its own interest and universal values such as freedom, equality and humanity, by obediently and blindly following the American administration' (Komšić 2007). The document uses expressions such as 'the most profound moral downfall in the history of mankind' and 'hypocrisy and neocolonialism' in condemning the Europeans (Komšić 2007: 28).

After 2008, the party was focused on Serbian EU integration, without having any publicly elaborated position on the EU. Its party programme only broadly stated that the party advocated the open borders in Europe that would allow for the free flow of people, goods, capital and ideas (SPS 2010). However, the interviewed senior party officials demonstrated a rather positive stance on this issue. The party vice president Dijana Vukomanović (2011) thus argued that the EU has proven to be the most convincing political, economic, cultural and civilizational model that unquestionably has no alternative. As a result, the party was classified as Soft Euroenthusiastic, since it appeared to belong to the middle category in the proposed concept, which describes a party without clearly formulated or expressed attitudes towards the EU, but with a tendency to adopt an affirmative position.

On the other side, the party has also undergone a major transformation over the last decade with regard to its position on Serbian EU accession. Although the party did not articulate outright rejection of Serbian EU membership, the Socialist Party of Serbia was widely perceived throughout as an unreformed and anti-European party. Party vice president Djukić Dejanović (2011) confirmed that 'the party has never been explicitly against Serbian EU integration', although 'its key policies may have expressed an anti-European orientation'. The party did not endorse a 2004 parliamentary resolution on Serbian accession to the EU, even though it had formulated the accession of Serbia to the EU as a political goal at the 2003 congress. Nevertheless, the party practically nullified this commitment through a decisive refusal to extradite those charged with war crimes to the ICTY (Goati 2009). The 2006 party programme that talked about 'one of the most important statesman of the 20th century, Slobodan Milošević' (SPS 2008: 2) was still consistent with its 1990s politics. The Socialists stood against any attempts to trade EU membership for the recognition of Kosovo, since 'the West has not given up "the carrot-and-stick policy" in its relation to Serbia' (SPS 2008: 44). However, they also argued that 'it is the obligation of Serbia to join the European community of nations in a way that it would secure its vital national interests and rapid economic development' (SPS 2008).

The key changes came after the death of its authoritarian leader Slobodan Milošević at the ICTY in 2006, when Ivica Dačić was elected as the new party president. Although a long-term devotee of president Milošević, Dačić realized that the party would have no future unless it embarked on a process of strategic re-positioning. He publicly argued that the Socialists would not be able to get a single vote in the future if stayed committed to the past, and that there was no place for party members who were not ready to understand the depth of party changes (*Blic* 2010). In the 2008 election campaign, the Socialists thus conducted a campaign oriented towards socioeconomic issues, while arguing that Kosovo must remain an integral part of Serbia at any cost. The party attitude towards the EU was rather vague and occasionally negative, given the support of main EU countries for Kosovo's independence, but it did not openly declare its position regarding the signing of the Stabilisation and Association Agreement. By adopting such policies, the party wanted to reconcile the seemingly irreconcilable – to keep its core national and anti-European voters, and at the same time to move closer to a pro-European pole. Its strategy was to play on the core electorate's dissatisfaction with the socioeconomic position, not on their national sentiments, which allowed for a gradual shift of party attitudes towards the pro-EU pole. Specifically, representation of this niche of the electorate's socioeconomic interests brought the party the opportunity to adopt affirmative attitudes towards the EU and allowed them to argue that only Serbian accession to the EU might improve their living standards. Its previous emphasis on nationalism, on the other hand, was naturally coupled with Euroscepticism.

The post-2008 election coalition building proved to be a decisive factor that reinforced the party's transformation. After successfully balancing between the apparent unwillingness of its electorate and some senior party members to accept a new policy orientation and the need to reinvent the party, Dačić made a highly strategic and pragmatic decision to help form a pro-EU government. This was a result of the self-interest of a new party leadership (Stojiljković 2007), which was trying to come closer to the social-democratic political family, as well as of a calculation that anti-European policies had absolutely no future (Bakić 2011). Djukić Dejanović (2011) confirmed that the change of party attitudes towards the EU was to a great extent the result of strategy and electoral tactics. She particularly stressed the experience of being in opposition when 'the Socialists were stigmatized and labelled as the Jews in Nazi Germany', as a factor that 'actually much helped the party to adopt European standards as its own'. Such a decision may also be attributed to the influence of the EU, which was interested in the formation of a pro-European government and directly interfered in the election process by supporting the pro-European coalition of Serbian president Boris Tadić. It is therefore reasonable to argue that the impact on the Socialists' decision to join the pro-EU government was important, although Djukić Dejanović (2011) claimed that 'there was no classical pressure from the EU. There were only conversations about the fact that it was worth supporting policy of president Tadić and the Democratic Party'.

New Serbia: Strategic 'Europeans'

The New Serbia (NS), characterized as a Soft Euroenthusiastic party, was founded as a monarchist and conservative party that stands for Serbia as the state of the Serbian people, the Serbian Orthodox Church as the backbone of moral and spiritual renewal, and a parliamentary monarchy (NS 2010). The programme of this party, unlike the large majority of Serbian parties, did briefly state what type of EU it supported. It specifically stated that the party's objective was not only to join the European institutions, but also to strengthen them by supporting the transformation of the European Union from 'commercial society' to 'political society' (NS 2010: 3). However, the party did not further elaborate this position and the issue of the EU was virtually absent from its rhetoric.

On the other side, the party has changed its attitude towards Serbian EU integration two times since 2008, and as such represents an unambiguous example of a party driven by strategic and tactical electoral considerations. The programme of the New Serbia (NS 2010: 3) demonstrated its European orientation, by stating that it believed that Serbia belonged to a United Europe not only geopolitically, but also because of its entire history and cultural heritage. It was part of the pro-EU coalition that ousted president Milošević and largely followed its traditional coalition partner – the Democratic Party of Serbia – in all key policies. Therefore, it supported Serbian EU membership until the 2008 Kosovo declaration of independence, when it took a strong Eurosceptic position and argued against Serbian accession to the EU. It opposed the signing of the Stabilisation and Association Agreement, accused the government of betraying national interests and announced its willingness to initiate the procedure for impeachment of the Serbian president Boris Tadić (NS 2011).

However, it again shifted attitudes towards Serbian EU membership in 2010, abandoned the coalition with the Democratic Party of Serbia and returned to its pre-2008 policy, due to a fear of political marginalization. This decision was solely pragmatic and strategic. The party vice president Filipovski (2011) confirmed that the voters' support had been in steady decline since 2008, and given that 'the voters have already made their decision' and that Kosovo was not a priority issue, the party decided to 'follow what the citizens of Serbia think' and return to its original pro-European principles. The party formed a coalition with the Serbian Progressive Party in the run-up to the 2012 elections and advocated EU membership, despite the fact that some senior party officials were 'emotionally deeply against it, given the negative consequences of EU policy in the Balkans' (Marić 2011). Therefore, this is one more example of pragmatic and strategic party positioning, since the party's Soft Euroenthusiasm (Table 6.2) is not deeply grounded and is devoid of principled beliefs or convictions.

Conclusion

The analysis of Serbian parties' positions on Europe through application of the proposed concepts (Table 6.2) points to several conclusions. First, the key line of division between the parties is whether Serbia should be an EU member state or not. Different views on the EU (both the EU as it is and the ideas underlying the process of European integration) are difficult to identify in parties' public rhetoric and are almost completely absent from their programmatic documents.

This can be explained by the fact that voters know almost nothing about the EU so political parties have no incentives to exploit these issues. In addition, the ignorance of the EU seems to be widespread in the parties themselves. As a result, this issue is neither the topic of political discussion nor the basis for party classification into Eurosceptic and Euroenthusiastic categories. It is therefore possible that the essentially pro-European Democratic Party of Serbia, which has not had a fundamentally negative attitude towards the EU, is characterized as Soft Eurosceptic, given its strong commitment to the political neutrality of Serbia.

Second, the study demonstrates that the principled pro- and anti-European parties, termed Hard Euroenthusiasts and Hard Eurosceptics, for the most part have adopted ideologically based attitudes to Europe, and have consistently opposed or supported both Serbian EU membership and the EU, although the latter is much less present in their public discourse. These parties have not changed attitudes since they were founded, although their rhetoric may have been occasionally and strategically moderated (the Serbian Radical Party in the mid-2000s). Specifically, the traditionalism versus modernism dimension of party competition seems to play an important role, given that party attitudes to Europe match their positions on this axis, with liberal and modernist parties being most pro-European, and traditional and national being rather Eurosceptic.

On the other hand, it appears that Soft Euroenthusiastic and Soft Eurosceptic parties, with the exception of the Democratic Party of Serbia, have adopted strategically motivated attitudes to Europe. As a consequence, they tend to change positions on Serbian EU membership (in the case of New Serbia even twice within two years), which indicates a lack of firm foundation of their attitudes. Moreover, these parties have not adopted clearly defined positions on the EU, and elements of both positive and negative stances on this issue may be identified. They often strategically choose not to speak about the EU, in order to conceal their rather critical position (especially in the case of the Serbian Progressive Party and the New Serbia) and to obtain 'the legitimacy of the West'.

Third, the study also identifies that party attitudes towards EU membership tend to shift more easily than their position on the EU. Serbian parties therefore have moved in the vertical dimension of the proposed concept, while horizontal changes have not occurred. This, coupled with a lack of elaborated positions, suggests that it is harder for parties to change attitudes and policies towards the EU and the ideas underlying the process of European integration, which thus appears to be to some extent more ideologically grounded and more revealing about parties' real positions on Europe. On the other hand, the question of joining the EU has proved to be a more practical issue on which parties responded strategically. This broadly confirms Kopecky and Mudde's (2002) arguments that ideology may determine a party's support for the ideas underlying the process of European integration, while the EU as it is (and EU membership in non-members) may be more strategically perceived.

The analysis also points to a specific position of the Democratic Party of Serbia, which is characterized as a Soft Eurosceptic party that is not ideologically anti-

European, but is distinctly against Serbian EU membership. This party position is unique in the way that it is not pragmatically and strategically driven (as expected from parties belonging to 'soft' categories), and is rather a consequence of a deep traditional and national ideological conviction that the international borders of European countries are unchangeable and that preservation of territorial integrity is a prime European value that is valued far more than Serbian EU membership. In line with Szczerbiak and Taggart's (2008) initial concept, this party has concerns about one of the EU policy areas, namely its policy towards Serbia, which de facto requires acceptance of Kosovo's independence for further EU integration. The party therefore argues that the national interest of preservation of Serbian territorial integrity is currently at odds with EU membership, making it a Soft Eurosceptic party.

Fourth, the largest number of Serbian parties belong to the 'soft' categories and oscillate between support and opposition to Serbia's EU membership, which points to a lack of deeply grounded views. There are no indications that, for example, the former Eurosceptics have adopted a pro-EU position due to a belief that Serbian EU membership has become intrinsically good. Their new stance is rather the result of rational political calculations and post-electoral negotiations influenced by the Western countries as well as repeated failed attempts to come to power by advocating nationalism and (sometimes rhetorically moderated) anti-Europeanism. In addition, the shifting nature of their attitudes shows difficulties in forming a clear and unequivocal position on Europe. This fact is certainly not a surprise given the complex relationship between the two sides over the past two decades. This also demonstrates that a wider social consensus on EU membership has not been reached. However, Table 6.3 also shows that the political significance of Eurosceptic parties has declined over time, particularly following the foundation of the Serbian Progressive Party in 2008, when all key parties started advocating Serbian EU membership. Further electoral decline of these parties was dramatically seen at the 2012 elections, when the Serbian Radical Party, once the largest party in the Parliament, failed to pass the electoral threshold and became a non-parliamentary party, while the Democratic Party of Serbia remained a small party without any possibility of implementing its political platforms.

Finally, the EU is almost exclusively perceived among the Serbian political elite through the prism of its policy towards Serbia and the former Yugoslavia over the past two decades. Specifically, (former) Eurosceptic parties' attitudes towards the EU and EU membership are directly related to the issue of Kosovo's status, as well as the policy of the EU and its leading Member States. That is certainly not a surprise, bearing in mind that there is no agreement among EU Member States about Serbian international borders, which is a unique case in Europe. In this way, the fundamental and unresolved issues of sovereignty, territorial integrity and borders resulting from the idiosyncratic, historical and political circumstances that arose from the violent breakup of the former Yugoslavia, decisively shape the politics of Serbian Eurosceptic (and not only Eurosceptic) parties towards the EU.

References

Bakić, J. 2011. Interview with the Author. Belgrade, April 2011.

Batory, A. 2008a. *The Politics of EU Accession: Ideology, Party Strategy and the European Question in Hungary.* Manchester: Manchester University Press.

Batory, A. 2008b. Euroscepticism in the Hungarian Party System: Voices from the Wilderness?, in *Opposing Europe? The Comparative Party Politics of Euroscepticism*, edited by A. Szczerbiak and P. Taggart. Vol. 1. Oxford: Oxford University Press, 263–76.

Blic 2010. Dačić: Policija nema dokaze da Šešelj stoji iza napada. Available at: www.blic.rs/Vesti/Politika/196845/Dacic-Kada-se-promeni-dirigent-pevaju-se-druge-pesme [accessed: 15 August 2010].

Cesid 2012. *Elections Results.* Available at: http://www.cesid.org/lt/ [accessed: 10 May 2010].

Conti, N. 2003. Party Attitudes to European Integration: A Longitudinal Analysis of the Italian Case. Susses European Insitutet – *SEI Working Paper*, No. 70.

Cvijanović, Z. 2011. Interview with the Author. Belgrade, April 2011.

Djukić Dejanović, S. 2011. Interview with the Author. Kragujevac, January 2011.

Djurković, M. 2011. Programski identiteti desnih partiji u Srbiji krajem prve decenije ovog veka, in *Partije i izbori u Srbiji – 20 godina*, edited by S. Orlović. Beograd: Friedrich Ebert Stiftung (Fakultet političkih nauka), 199–218.

DSS 2010a. *Party Statement.* Available at: www.dss.rs/newsitem.php?id=9063 [accessed: 25 September 2010].

DSS 2010b. *Party Statement.* Available at: www.dss.org.rs/newsitem.php?id=9638 [accessed: 25 September 2010].

DSS 2010c. *Party statement.* Available at: www.dss.org.rs/newsitem.php?id=9471 [accessed: 25 September 2010].

DSS 2012. *Party programme.* Available at: http://dss3.orion.rs/wp-content/uploads/downloads/2011/02/04-Program-%E2%80%93-Demokratske-stranke-Srbije.pdf [accessed: 1 April 2012].

Emerson, P. and Šedo, J. 2010. Electoral Systems and the Link to Party Systems, in *Party Politics in the Western Balkans*, edited by V. Stojarova and P. Emerson. Oxon: Routledge, 6–25.

Fallend, F. 2008. Euroscepticism in Austrian Political Parties: Ideologically Rooted or Strategically Motivated?, in *Opposing Europe? The Comparative Party Politics of Euroscepticism*, edited by A. Szczerbiak and P. Taggart. Vol. 1. Oxford: Oxford University Press, 201–20.

Filipovski, D. 2011. Interview with the Author. Belgrade, February 2011.

Goati, V. 2009. State Interest through the Prism of the Commitments of Ruling Parties, in *Between Authoritarianism and Democracy: Serbia at the Political*

Crossroads, edited by D. Vujadinović and V. Goati. Belgrade: CEDET/FES, 283–302.

Government of the Republic of Serbia 2010. *An Address of Vojislav Koštunica*. Available at: www.arhiva.srbija.gov.rs/vesti/2004-03/02/343974.html [accessed: 18 July 2010].

Henderson, K. 2008. The Slovak Republic: Eurosceptics and Phoney Europhiles, in *Opposing Europe? The Comparative Party Politics of Euroscepticism*, edited by A. Szczerbiak and P. Taggart. Vol. 1 Oxford: Oxford University Press, 277–94.

Komšić, J. 2007. Političke stranke u Srbiji i evropske vrednosti: Programi i praksa, in *Političke Stranke u Srbiji i Evropska Unija*, edited by Z. Lutovac. Beograd: Friedrich Ebert Stiftung (Fakultet političkih nauka), 9–49.

Kopecky, P. and Mudde, C. 2002. The Two Sides of Euroscepticism: Party Positions on European Integration in Central Eastern Europe. *European Union Politics*, 3(3), 297–326.

Marić, J. 2011. Interview with the Author. Belgrade, March 2011.

Martinović, A. 2011. Interview with the Author. Belgrade, April 2011.

Mirović, D. 2011. Interview with the Author. Belgrade, March 2011.

National Assembly of the Republic of Serbia 2012. *Assembly Composition*. Available at: www.parlament.gov.rs/narodna-skupstina-/sastav/politicke-stranke.904.html [accessed: 10 April 2012].

NS 2010. *Party Programme*. Available at: www.nova-srbija.org/index1.htm [accessed: 20 October 2010].

NS 2011. *Party Statement*. Available at: www.nova-srbija.org/index1.htm [accessed: 7 November 2010].

Parliamentary Network 2010. *An address of Vojislav Koštunica*. Available at: www.skupstinskamreza.rs/portal/index.php/politika/intervju/5316 [accessed: 14 August 2010].

RFERL 2007a. News. [Online: Radio Free Europe Radio Liberty]. Available at: www.rferl.org/content/article/1076353.html [accessed: 20 March 2012].

RFERL 2007b. News. [Online: Radio Free Europe Radio Liberty]. Available at: www.rferl.org/content/article/1076353.html [accessed: 20 March 2012].

Samardžić, S. 2011. Interview with the Author. Belgrade, March 2011.

SEIO 2012. *Serbia's EU Integration Supported by 51% of Citizens*. [Online: Serbian European Integration Office]. Available at: www.seio.gov.rs/news.101.html?newsid=1121 [accessed: 15 March 2012].

Siaroff, A. 2000. *Comparative European Party Systems: An Analysis of Parliamentary Elections Since 1945*. New York, NY: Garland Publishing.

SNS 2010a. *Programmatic Principles*. Available at: http://sns.org.rs/%D1%81%D1%80/2010-06-0-00-1-55/srpska-napredna-stranka-principi-delovanja.html [accessed: 20 July 2010].

SNS 2010b. *Party Statement*. Available at: http://sns.org.rs/%D1%81%D1%80/srpska-napredna-stranka-u-medijima/112-2010-06-04-22-17-07/2541--srpska-napredna-stranka-jul.html [accessed: 25 August 2010].

SNS 2010c. *Party Statement*. Available at: http://sns.org.rs/%D1%81%D1%80/srpska-napredna-straka-vesti/85-glavne-vesti/2407--srpska-napredna-stranka-jun.html [accessed: 25 August 2010].

SNS 2011a. *Party News*. Available at: http://www.sns.org.rs/%D1%81%D1%80/srpska-napredna-straka-vesti/85-glavne-vesti/3367--srpska-napredna-stranka-tanjug.html [accessed: 25 August 2010].

SNS 2011b. *Party Programme*. Available at: http://sns.org.rs/images/pdf/Bela%20knjiga-Programom-do-promena.pdf [accessed: 10 December 2011].

SNS-FPÖ 2011. *Cooperation Agreement*. Available at: http://sns.org.rs/images/stories/pdf/Sporazum-o-partnerstvu-SNS-i-Slobodarske-partije-Austrije.jpg [accessed: 12 August 2011].

SPS 2008. *Old Party Programme*. Available at: www.sps.org.rs/sr/dokumenta/program.html [accessed: 20 July 2010].

SPS 2010. *Party Programme*. Available at: http://sps.org.rs/Dokumenta/PROGRAM%20SPS.pdf [accessed: 10 December 2011].

SRS 2009. *Party Programme*. Available at: www.srpskaradikalnastranka.org.rs/pdf/misc/20091023-program.pdf [accessed: 15 August 2010].

SRS 2010. *Party Statement*. Available at: www.srpskaradikalnastranka.org.rs/srbija/1415 [accessed: 2 November 2010].

Stojić, M. 2010. Europe and the Serbian Parliamentary Election of May 2008. *EPERN Election Briefing No. 50*.

Stojiljković, Z. 2007. Socijaldemokratija i političke stranke Srbije, in *Ideologija i Političke Stranke u Srbiji,* edited by Z. Lutovac. Beograd: Friedrich Ebert Stiftung (Fakultet političkih nauka), 113–54.

Szczerbiak, A. 2008. Opposing Europe or Problematising Europe? Euroscepticism and "Eurorealism" in the Polish Party System, in *Opposing Europe? The Comparative Party Politics of Euroscepticism*, edited by A. Szczerbiak and P. Taggart. Vol. 1. Oxford: Oxford University Press, 221–42.

Szczerbiak, A. and Taggart, P. 2008. Introduction: Researching Euroscepticism in European Party Systems: A Comparative and Theoretical Research Agenda, in *Opposing Europe? The Comparative Party Politics of Euroscepticism*, edited by A. Szczerbiak and P. Taggart. Vol. 2. Oxford: Oxford University Press, 2–27.

Velika Srbija 2011. We will not go to the EU. *Velika Srbija: Bulletin of the Serbian Radical Party*, October 2011, 3475, 7–8.

Vukomanović, D. 2011. Interview with the Author. Belgrade, March 2011.

Chapter 7

Bosnia and Herzegovina in the EU Integration Process: A Carrot-and-Stick Marathon

Emir Vajzović

The enlargement relationship between the European Union and Western Balkan countries is defined by a carrot-and-stick approach. Although broadly similar for every country, there are specific differences in the process for each of the countries that emerged after the dissolution of the former Yugoslavia. Bosnia and Herzegovina is of special interest in this regard in view of its troubled recent history, including the 1992–1995 war, its specific past, worried present, and questionable future. However, in order to better understand today's complex relations in the region and within Bosnia we need to review the main historical events that led to the current situation. Today, Bosnia and Herzegovina is going through numerous transformations, reforms, consolidation and integration processes, with the EU playing a significant role in all of them. The government of Bosnia and Herzegovina and its citizens decided long ago that they see their future within the EU as the only foreseeable solution for multiple reasons: political stabilization, financial assistance, security and hope for a better tomorrow. This chapter provides a historical introduction and analysis of the events that led to today's relationships in Bosnia and Herzegovina as well as to the post-war involvement of the EU in Bosnia and Herzegovina and its balancing of potential reward (i.e. EU financial assistance and potential membership seen as *l'aboutissement du rêve* [the dream come true/solution for all problems]) with the threat of punishment (i.e. being left behind in the Western Balkans EU integration process as the only (unprotected) isolated land) to induce behaviour. Furthermore, the EU's presence in Bosnia and Herzegovina is far more obvious and specific than in other ex-Yugoslav republics. Bosnia and Herzegovina is practically a semi-protectorate, with the presence of the Office of the High Representative (OHR) and EU Special Representative (EUSR), along with police and military missions (EUPM and EUFOR) and strong direct political influence on the political elites. Even so, the process of EU-led democratic consolidation and institutional (capacity) building has been slow and painful. Why?

Beginning of the Marathon: The Starting Line

For some, the dissolution of the Socialist Federal Republic of Yugoslavia came as a total surprise, while for others it was seen as an inevitability with (un)foreseeable elements and forms of dissolution. The armed conflict that took place in various forms (from a small conflict (Slovenia) to genocide (in Bosnia and Herzegovina)) has set back the democratic transition processes that have just started. In the post-Cold War era, former Yugoslav republics tried to overcome the overwhelming challenges they faced. In our case, we will look at Bosnia and Herzegovina, which was confronted by probably the most challenging tasks. These included multiple transition processes, post-war reconciliation and reconstruction on various levels, including infrastructure, political and social institution building, and so on. In addition, there were the issues of the international community's semi-protectorate and the European Union's pragmatic and sometimes cynical approach to integration.

The current situation in Bosnia and Herzegovina will be presented through several key moments, starting with the shaping events – the war (1992–1995), initiated by the dissolution of the Yugoslav federation and ended with the General Framework Agreement for Peace in Bosnia in Herzegovina, known as the Dayton Peace Agreement (DPA), under the patronage of the international community. The DPA acknowledged the current status quo – front lines became entity borders and conflicting sides became the actors in political reality, with the Constitution of Bosnia and Herzegovina (Annex IV of DPA) as the minimum requirement for state existence and the maximum that parties to the Peace Agreement could accept. As Pejanović (2010: 79) observes:

> the general idea of this plan (DPA) was to stop the war at separation lines between two territorial units established during the war: Federation of Bosnia and Herzegovina as unit established by the Washington Peace Treaty and Republic of Srpska (as the result of ethnic cleansing of non-Serb population) with respective territorial parts of 51 per cent and 49 per cent. Such solution has favoured ethnic criteria trough ethnic division and ethnic territory allocation in Bosnia and Herzegovina. Therefore, ethnic criterion has become the dominant element of internal constitutional framework of Bosnia and Herzegovina.

To understand the current challenges facing Bosnia and Herzegovina (BiH), it is essential to offer an overview of the historical context that has largely determined and shaped current relations. Modern ethno-political elites in BiH and elsewhere in the region are prone to refashioning the complex history of the region in terms of the fear of others in order to build a strong electoral base. Many authors (both local and foreign) have made their contribution to preserving the picturesque history of Bosnia and Herzegovina and offer it to a wider audience. Malcome's (2011) retrospective of Bosnia's history offers an overview with analysis of recent events. Furthermore, Kent (1999) has also summarized the very complex history of Bosnia and Herzegovina, explaining that, although it is smaller than West Virginia or Ireland, Bosnia and Herzegovina has long been a meeting ground and crossroads for people of different cultures and different empires. How did it start? We will not go into the detailed history, but a few remarks are needed to understand better what is happening today in Bosnia and Herzegovina and the analysis made later in the text.

The Slavic ancestors of today's inhabitants moved south into the Balkan Peninsula in the sixth and seventh centuries AD. They intermixed with the people who already lived in this borderland between the Roman and the Byzantine empires. Both Rome and Constantinople sent Christian missionaries to the region. Some Bosnians chose Roman Catholicism as a religion, and others chose Eastern Orthodoxy. Over time, the Roman Catholics came to see themselves as Croats. They wrote in the Latin alphabet and identified with Croatian culture. The Eastern Orthodox, however, came to see themselves as Serbs. They wrote in the Cyrillic alphabet and identified with Serbian culture. Thus, for the first centuries of the Christian era, Bosnia was part of the Roman Empire. In the ninth century, the kingdoms of Serbia and Croatia shared control of Bosnia. This was followed in the eleventh and twelfth centuries by rule of the region by the kingdom of Hungary. The medieval kingdom of Bosnia gained its independence around 1200 and Bosnia remained independent until 1463, when the Ottoman Turks conquered the region. During Ottoman rule, many Bosnians converted from Christianity to Islam (some of them were later seen as Muslims (nationality)/Bosniaks). Bosnia was under Ottoman rule until 1878, when the Congress of Berlin transferred administrative control to Austria–Hungary and annexed Bosnia in 1908.

While those living in Bosnia came under the rule of the Austro–Hungarian Empire, South Slavs in Serbia and elsewhere were calling for a South Slav state. The trigger for the First World War was when Serb nationalist Gavrilo Princip assassinated the Archduke Franz Ferdinand in Sarajevo. After the War, Bosnia became part of the South Slav state of Yugoslavia, only to be given to the Nazi-puppet state, the Independent State of Croatia in the Second World War (WWII). During this period, many atrocities were committed against Jews, Serbs, and others who resisted the occupation. During and after WWII, under Josip Broz Tito, the state of South Slaves was created: the Socialist (Communist) Federal Republic of Yugoslavia, with Bosnia re-established as a republic with its medieval borders within the Yugoslav federation.

For the purpose of strengthening their position of power in the region, throughout the last 200 years Bosnia was under heavy influence and interference from Croatia and Serbia, which were slowly but surely homogenizing people along religious/ethnic lines (Croatia – all Catholics in the region should be of Croat national identity with Croatia as their home state and seat of political power, and similarly with Serbia and the Orthodox in the region). However, throughout history and under various empires and periods of independence, Bosnia (and Herzegovina) has managed to preserve its multinational and multi-religious character, as well as the stable elements of statehood (Smajić 2011: 66).

Nevertheless, this Bosnian character 'united in diversity', which was built up throughout history, was quickly turned into 'deeply divided society' by the ethno-political elites and the international community in Bosnia and Herzegovina from the beginning of the 1990s, in which some (or most) Bosnian Croats and Serbs feel as if they have a second/original/alternative/mother state, that is the Republic of Croatia and Republic of Serbia, respectively. This had (and still does have) a significant influence on events in the 1990s and on the current political situation in Bosnia and Herzegovina.

The US Department of State in its *Background Note on Bosnia and Herzegovina* (2012) addresses recent events that have shaped the contemporary Western Balkans:

> Yugoslavia's unravelling was hastened by Slobodan Milošević's rise to power in 1986. Milošević's embrace of Serb nationalism led to intrastate ethnic strife. Slovenia and Croatia both declared independence from Yugoslavia in June 1991. By late September 1991, Bosnian Serb Radovan Karadžić's SDS had declared four self-proclaimed "Serb Autonomous Regions (SAO)" in Bosnia. In October 1991, the Bosnian Serbs announced the formation within Bosnia of a "Serbian Republic of Bosnia-Herzegovina" that would have its own constitution and parliamentary assembly. In January 1992, Radovan Karadžić publicly proclaimed a fully independent "Republic of the Serbian People in Bosnia-Herzegovina". On March 1, 1992, the Government of Bosnia and Herzegovina held a referendum on independence from SFR Yugoslavia. Bosnia's Parliament declared the Republic's independence on April 5, 1992. However, this move was opposed by Serb representatives, who had voted in their own referendum in November 1991 in favour of remaining in Yugoslavia.

Bosnian Serbs, supported by neighbouring Serbia and the Yugoslav National Army, responded with armed force in an effort to partition the Republic along ethnic lines to create a 'greater Serbia'. Full recognition of Bosnia and Herzegovina's independence by the United States and most European countries occurred in April 1992, and Bosnia and Herzegovina was admitted to the United Nations on 22 May 1992.

Radical elements in Croatia led by President Franjo Tudjman saw the opportunity to take part of Bosnia and Herzegovina as well, opening a new front in Herzegovina. Hoare (1997) argues that Croatian policy in Bosnia was governed by the ebb and flow of the events of 1991–1993, above all because, unlike its Serbian counterpart, the Croatian project to partition Bosnia had little ideological basis to pre-determine the form of its implementation. It was viewed by its proponents as a possible, not as a necessary option, to be realized if events required and demanded (Hoare 1997). Croatian national ideologues traditionally viewed the whole of Bosnia and Herzegovina as a Croatian land and Bosnia's Muslims as 'Croats of the Islamic faith'. Tudjman himself subscribed to this tradition and, together with Milošević, instigated an 'agreed war' intended to strengthen two nationally homogeneous countries – Croatia and Serbia (Malcom 2011: 374–5, Smajić 2011: 73–86).

By March 1994 (the Washington Agreement), Muslims (Bosniaks) and Croats signed an agreement creating the Federation of Bosnia and Herzegovina. This narrowed the field of warring parties to two. The conflict continued through most of 1995, and many atrocities were committed, including acts of genocide committed by the Army of Republika Srpska in and around Srebrenica between 12 and 22 July 1995, where approximately 8,000 Bosnian Muslim men and boys were killed (European Parliament 2009).

In order to bring justice and ensure that such events would not be repeated anywhere else, on 25 May 1993 the United Nations Security Council passed Resolution 827 establishing the International Tribunal for the Prosecution of Persons Responsible for Serious Violations of International Humanitarian Law Committed in the Territory of the Former Yugoslavia since 1991 (ICTY). Despite its uneven performance, since 1993, the ICTY has irreversibly changed the landscape of international humanitarian law and provided victims with a voice. In its precedent-setting decisions on genocide, war crimes and crimes against humanity, the Tribunal has shown that an individual's senior position can no longer protect them from prosecution. Unfortunately, ICTY has done little to bring truth and reconciliation to the region, and Bosnia and Herzegovina, in particular.

The role of post-war reconciliation and state building has been the responsibility of the international community (predominantly the USA and the EU). In what follows, we will review how the international community, and especially the EU, have applied a carrot-and-stick approach and to what extent it was successful. The carrot paradigm refers to a policy of offering a combination of rewards and punishment to induce positive behaviour, thus the use of promises of reward and threats of punishment to make somebody work harder and meet necessary conditions.

The Post-Dayton Era

Bosnia and Herzegovina was not the same before and after the Dayton Peace Agreement and this point is often used to determine the beginning of a new epoch with apparent peace (actually more an absence of armed conflict) and painstaking reconstruction (infrastructure, society, democracy, economy, hearts and minds).

The USA-led peace negotiations at Wright-Patterson Air Force Base in Dayton, Ohio, mediated by the late Ambassador Richard Holbrooke, resulted in a peace agreement, ending the war on 21 November 1995 (Holbrooke 1998). Bosnia and Herzegovina today consists of two Entities: the Federation of Bosnia and Herzegovina, with a majority of Bosniaks and Croats, and the Republika Srpska, which is primarily Serb. This is the result of the international community's acceptance of the division of Bosnia and Herzegovina along ethnic lines and the war-generated administrative boundaries and ethnic cleansing (massive-scale internal forced migrations aimed towards ethnically 'clean' territories) (Ibrahimagić 2010b: 298–9) along with a post-war policy of preserving such a demographic situation (Lavić 2012: 75–9, Petrović 2012: 53–63).

The organizational units in Bosnia and Herzegovina are the following: at State level there are *Institutions of the State*, with a bicameral Parliament Assembly of Bosnia and Herzegovina (House of Representatives and House of Peoples), directly elected Presidency of Bosnia and Herzegovina (rotating presidency with three presidents (Bosniak, Croat and Serb) and Council of Ministers headed by a Chairman, along with a Constitutional Court and Central Bank. State institutions have limited powers since more jurisdictions are preserved by the Entities (the Federation of Bosnia and Herzegovina (divided into ten cantons) and the Republic of Srpska). As Article III (3a) of the Bosnia and Herzegovina Constitution reads: 'All governmental functions and powers not expressly assigned in this Constitution to the institutions of Bosnia and Herzegovina shall be those of the Entities.' The following matters are the responsibility of the institutions of Bosnia and Herzegovina (according to the DPA): (a) foreign policy; (b) foreign trade policy; (c) customs policy; (d) monetary policy as provided in Article VII; (e) finances of the institutions and for the international obligations of Bosnia and Herzegovina; (f) immigration, refugee, and asylum policy and regulation; (g) international and inter-Entity criminal law enforcement, including relations with Interpol; (h) establishment and operation of common and international communications facilities; (i) regulation of inter-Entity transportation; and (j) air traffic control. Furthermore, one of the success stories is the reform of the defence sector and single armed forces (Aybet and Bieber 2011).

Entities have their own constitutions as well as legislative, executive and judicial institutions and often act as states within the State. The City of Brčko was left by the DPA to be determined by international arbitration and was later proclaimed the District of Brčko, independent from the other two Entities. There are three main ethnic groups in Bosnia and Herzegovina: Bosniaks, Serbs and Croats (or the 'Constitutive Peoples' as indicated in the Constitution), and languages in use are Bosnian, Serbian, and Croatian (all formerly known as 'Serbo-Croatian'). The ethno-political elites in Bosnia and Herzegovina are trying to affirm Ethnic groups as National groups, with emphasis on differences and strengthening elements of national identity (common territory, history, culture, economy and often common language – as minimum requirements – plus the common enemy – all others). Definition of nations in Bosnia and Herzegovina depends on subjective and psychological variables as opposed to objective ones. To be Bosniak/Croat/Serb is more a 'feeling' than an objective reality (Vajzović 2004: 20). However, these ethno-political elites have now been in place for almost 20 years deepening differences, and the international community has done little to change this (Domm 2007, 2011).

In truth, all three languages constitute a single, mutually comprehensible, language. However, from one language diasystem, three separate national standard languages were developed and 'reserved' for their respective Constitutive Peoples, based more on political decisions and less upon linguistic needs (Vajzović 2008). The Nationalities/Constitutive Peoples are Bosniak (Muslim), Bosnian Serb, and Bosnian Croat, each respectively identified religiously with Islam, Serb Orthodoxy, and Roman Catholicism.[1] Even though many other ethnic groups have shaped the country's history, they were not recognized in the Dayton Constitution. The Preamble of the Bosnia and Herzegovina Constitution reads: 'Bosniaks, Croats, and Serbs, as constituent peoples (along with Others), and citizens of Bosnia and Herzegovina hereby determine that the Constitution of Bosnia and Herzegovina.' It is not clear what or who are the 'Others' and 'citizens' (separately stated) and what is meant by Bosnians and Herzegovinians (Abazović et al. 2010). Often, 'Others' is interpreted as national minorities. The 2003 Law on the Protection of Rights of Members of National Minorities states that BiH will protect the status, equality and rights of the 17 national minorities present in Bosnia and Herzegovina: Albanians, Montenegrins, Czechs, Italians, Jews, Hungarians, Macedonians, Germans, Poles, Roma, Romanians, Russians, Ruthenians, Slovaks, Slovenians, Turks and Ukrainians. However, the ethno-political elites present continuous obstructions in order to persuade these minorities to choose a Bosnian-Herzegovinian national identity.

In July 2000, the Constitutional Court of Bosnia and Herzegovina reached a decision whereby Bosniaks, Croats and Serbs would be recognized as constituent people throughout the territory of Bosnia and Herzegovina. This constituted systematic discrimination in the country. In March 2002, this decision was formally recognized and agreed by the major political parties in both Entities.

Nonetheless, the main challenge for Bosnia and Herzegovina is its constitutional organization: is it a Federation, Confederation, something in between or neither? Ibrahimagić views the problem not as one of constitutional disagreement but rather as a consequence of imperialism of Bosnia and Herzegovina's neighbours and local ethnonationalism:

> Serbia, Montenegro and Croatia have started aggressive war against Bosnia and Herzegovina with intention of territorial expansion on account of state territory of Bosnia and Herzegovina recognized by international law. They have instructed their fellows – Serbs, Croats and Montenegrins which are living in Bosnia and Herzegovina for sake of implementation of their greater state policies by providing arms and turning them against their homeland – Bosnia and Herzegovina, by creating parastatal national entities, some of which are legitimised by international law – the Federation of Bosnia and Herzegovina by Washington Treaty (18 March 1994) and Republic of Srpska by Dayton Agreement (21 November 1995) signed on 14 December 1995. (Ibrahimagić 2010b: 13)

---

1 According to the 1991 census, the total population was 4,377,033 (Bosniak 43.47 per cent, Serb 31.21 per cent, Croat 17.38 per cent, Others 7.94 per cent).

In order to ensure stability in the region, even 20 years after its independence and the beginning of war in Bosnia and Herzegovina, the international community maintains an extraordinary civilian and military presence in Bosnia and Herzegovina stemming from the Dayton Peace Accords (now EUFOR and EUPM). However, it is to be seen whether this presence has well-balanced quantitative and qualitative engagements. How effective has this engagement been?

From Dayton to the Brussels Era

It is often said that Bosnia and Herzegovina is undergoing difficult transitional reform processes and it is therefore logical to expect obstacles in their implementation. More specifically, it is enduring a (at least) quadruple transition: from single-party system to multi-party democracy, from self-management to a market economy, from war to peace and reconciliation, and from an ethnically divided and segregated country as a result of war policy and ethnic cleansing to a more integrated and functioning state. In addition to this, or as the symbolic end-point of this transitional process, the country is expected to become a member state of the European Union in the foreseeable future. This has been characterized as the 'Dayton to Brussels' transition by some observers (Aybet and Bieber 2011, Coffey International Development 2009).

The Dayton Peace Agreement created the Office of the High Representative, an international institution to oversee implementation of the civilian aspects of the agreement, with powers to impose laws and revoke elected officials acting against Dayton. Thereby the form of semi-protectorate (or full-scale protectorate, as some argue) is enforced along with the introduction of democratic mechanisms in Bosnian society. However, the society and citizens of Bosnia and Herzegovina were unprepared for such a difficult and sudden transition without a previously built democratic culture and values and disturbed by three-and-a-half years of bloody war.

The international community has, metaphorically speaking, sought to have Bosnia and Herzegovina society learn how to swim in democratic water through a learning-by-doing method: enter the water and eventually you will learn to swim (or to put it more colloquially: democracy is imposed – deal with it!). One of the problems was that the DPA and its Annex IV – the Constitution of Bosnia and Herzegovina – left Bosnia and Herzegovina society 'tied hand and foot' and it was pushed into the unpredictable harsh 'democratic sea' to learn how to swim, while there are still others in the water that would not mind seeing Bosnia and Herzegovina drown. Bosnia got very tired struggling to swim forward (still tied hand and foot) while the trainer (the EU) offered 'carrots' (which are difficult to consume while stressed to keep one's head above water) and threatened with the 'stick', and in fact did not provide any usable coaching advice other than: *If you want to win, you should not lose.*

Some questions will be easier to answer by looking at the administrative and political organization of Bosnia and Herzegovina today. Institutional organization goes beyond a regular federal or confederated state and looks more like an 'asymmetric confederation' (Kasapović 2005: 4). Some mechanisms of consociated democracy are integrated in the DPA and Constitution, but instead of providing solutions and protection measures, they are more often mechanisms of obstruction and further weakening of the state. The so-called 'entity vote' or 'entity veto' mechanism in Parliament is intended to protect group interests (of constitutive peoples); instead it is often misused (Bieber and Keil 2009, Kunrath 2010). Since the post-war political reality in Bosnia and Herzegovina has not given room for too much optimism, hope was placed in the international community and integration processes that were under way on the European continent.

The transition from former socialist regimes and post-conflict reconstruction, along with democratic consolidation, have turned the Western Balkan countries towards EU integration as the logical regional response to globalization – specially the ex-Yugoslav republics, which need a strong stimulant to go from arms to 'carrots', and thereby ensure the stability and prosperity of the region.

Since the birth of the idea, the European Union has grown from six to 27 Member States as a result of a series of enlargements. It was the fifth enlargement in two successive waves, in 2004 and 2007, to include 12 new Member States that defined the contours of the enlargement policy. This now covers the countries applying for EU membership (under the enlargement process) and the potential candidates of the Western Balkans (under the Stabilisation and Association Process). The main framework for enlargement is set out in Article 49 of the Treaty on European Union and the Copenhagen criteria. Official EU reasoning is that (a) the objective of the enlargement process is to prepare the applicant countries so that they can assume their obligations as Member States on accession, and (b) the Stabilisation and Association Process aims gradually to bring the potential candidate countries closer to the EU. These processes are based on strict conditions, with due regard for the specific needs and merits of each country, and are supported by bilateral and financial instruments devised for this purpose.

EU conditionality is a strategy of reinforcement by reward with set criteria and conditions with visions of a better future, prosperity, security and cooperation in the region. With substantial funds introduced through different instruments and programmes, along with decentralized project management and enabling and empowering domestic institutions to carry out the process of transition and European integration, the EU is slowly but surely achieving its goal of regional stabilization and security, ensuring that its future outer borders are safe and boosting the EU economy, thereby satisfying all criteria (political, economic and legal) – preserving its values of democracy and human rights along with strengthening its position in the global environment as a rising leader, primarily regionally but also on the world stage.

But can the same methodology be applied in all cases? Bosnia and Herzegovina is definitely not an 'ordinary' state. For example, on one hand, there is Croatia, a prospective member which faces no major difficulties in meeting the political requirements for membership, and Serbia, as a geopolitically interesting candidate for the EU (Stahl 2011), and on the other hand, there is Bosnia and Herzegovina as a potential candidate that has tried to make considerable progress in order to secure candidate status, but its complex constitutional structures and nationalism often lead to blockages and inefficiency in decision making (Freyburg and Richter 2010).

The European Commission regularly reaffirms that each country will advance towards this goal on its own merits, depending on its progress in meeting the requirements. However, these requirements are not always clear and firm. The countries are at different stages along their path towards the EU, but all will be able to realize their European perspective by following the road map – at least this is the official line. To assist them in this journey, the EU has provided throughout the years different financial instruments with significant budgets to support reform processes and to induce stimulation for the accession process.

As a very attractive carrot, from 2000 to 2013, the EU allocated to Bosnia and Herzegovina a total of 1.1 billion euros. What value the beneficiary countries and institutions realistically received and how effectively the funds were used by beneficiary countries cannot be discussed here, though both are questions worthy of further research. However, the question remains whether the different incentives granted to Croatia or Serbia and Bosnia and Herzegovina resulted in different levels of development and strengths of institutions and have contributed to the current different status in their EU integration. It appears that EU carrots were more efficient and sweet and that Western Balkan countries cannot get enough, regardless of the size and shape of the stick.

Regarding sticks, the conditionality approach of the EU to Bosnia and Herzegovina, among other requirements, first and foremost is constitutional reform with the goal of ensuring a functional democracy and equality and non-discrimination for all citizens of Bosnia and Herzegovina. Just as a reminder, the current Constitution of Bosnia and Herzegovina is in Annex IV of the Dayton Peace Agreement – a peace treaty document intended to stop the war (and not a long-term solution). The Dayton Peace Agreement is widely criticized for its content and implementation methods. Mujkić states (2008) that Dayton's Bosnia and Herzegovina has proven to be an inefficient creation which relies on generating crises from which the power is used by ethno-nationalist elites, which fundamentally relies on discrimination of citizens. What has the EU done with its 'stick', and how and why is it involved in Bosnia and Herzegovina? What is the difference in approach of the EU between Croatia and Serbia on one side, and Bosnia and Herzegovina on the other?

EU Involvement in Bosnia and Herzegovina

With the fall of the Berlin Wall, a wave of political changes came also to the Balkans. As a legacy of almost 50 years of communism in the region, people were more susceptible to continuing to cherish the cult of the leader, even in multi-party systems. Some of those leaders brought independence and 'democratic' mechanisms to their countries but remained as key political figures who suppressed development of a democratic culture.

Since the wartime leaders have been replaced by more democratic politicians, the Balkan states have attempted to build more democratic governments and societies. However, the main dilemmas relate to security and unresolved statehood. This has dominated political life and influences how the European Union interacts with the region. Fears of a chain reaction in the Balkans' 'powder keg' (i.e. the re-instigation of movements to create 'greater' nations in order to enlarge the boundaries of the state beyond historical boundaries, beyond those areas where the national ethnic group was in the majority or to adopt ethnic cleansing as a political doctrine), and that this could spread to the rest of Europe, continue to define the EU's engagement policy and the ways in which the accession and pre-accession agenda is articulated. As a result, the stability interests of the region are sometimes prioritized over other considerations, including the consolidation of democracy. According to Balfour and Stratulat, '[t]his approach undervalues a crucial lesson learned from European integration and enlargement: democracies are the most suitable polities to solve security and state-building problems, but not a perfect solution. At the heart of the problem is the state: its weakness is a major challenge for carrying out the necessary reforms for integration into the EU and for the resolution of internal and neighbourhood problems' (Balfour and Stratulat 2011: 1).

Following a period of confusion in EU foreign policy towards Bosnia and Herzegovina (e.g. its second-row seat (behind the USA) in mediation, security stabilization and post-war reconstruction and consolidation), with the beginning of the 2000s the EU took up more of a presence, politically and financially in the country (Džihić 2010). As a result of the increased political and (decreased) military and police presence, financial assistance grew progressively as well. Both 'sticks' and 'carrots' were increasingly introduced.

Although scaling down the US engagement in the Balkans after 11 September 2001 is consistent with the process of European integration, now seen as 'the only game in town', the EU should encourage that engagement to continue, especially in view of the fact that the United States enjoys strong credibility in the region (particularly in Kosovo and Bosnia) and that its professed primary goal, precisely, is to assist the region's accession to the EU, or in short: European integration strengthened by an Atlantic insurance policy (Rupnik 2011: 8).

The first post-war period in Bosnia and Herzegovina (from 1996 to the 2000 elections) was characterized by regrouping, determining parallel truths in public discourse and infrastructure reconstruction with the assistance of the international community. After the war, the EU started to become more involved in the region and Bosnia and Herzegovina. It started slowly, first in 1998 when the EU/Bosnia and Herzegovina Consultative Task Force was established. In 1999 the EU proposed the Stabilisation and Association Process (SAP) for five countries of South Eastern Europe, including Bosnia and Herzegovina. In June 2000 the Feira European Council stated that all the SAP countries were potential candidates for EU membership. In November 2000, the Zagreb Summit launched the SAP for those five countries. At that time, EU officials were still talking about a 'regional approach', hoping that the former Yugoslav republics might all be approaching the EU at the same pace. The 'regional approach' is to be understood here as the intention by the EU to integrate several countries of the region at the same time.

In adopting the 'regional approach' towards the Western Balkan countries, the EU's aim was to reconcile and rehabilitate relations between countries by introducing European values and standards, such as democracy and the rule of law, in order to foster their transition to a peaceful, stable and prosperous region. This aim was strengthened for the purpose primarily of securing and stabilizing the region by offering incentives that would draw the countries closer to the EU politically and economically. Although the initial aim of the EU regional approach was to deal with the countries as a 'group' or in a 'package', the particularities of each country were taken into account so that the assessment of their performances was based on their individual merits (Žarin 2007: 514). Soon it was clear that state-building and multi-stage transitional processes were going at different speeds, therefore this regional approach in terms of timing was soon abandoned.

According to Misita, 'in mid-2007 Bosnia and Herzegovina was characterised by two phenomenons: national-political outsmarting by political establishment in two entities from one side and insufficiently determined "arm twisting" by so called international actors, from the other side' (Misita 2007: 139). Some 'arm-twisting', or the use of the stick, as we call it, should obviously have been used with more determination, since both the opportunity and the need existed. However, as a consequence of these initial processes of EU integration in the region, Bosnia and Herzegovina has been a potential candidate for EU accession since the Thessaloniki European Council in June 2003 and has made use of the EU autonomous trade measures since 2000. It has been seen that the enlargement process was considered to be the best catalyst for political and democratic (unstoppable) transformation in the region as a whole. Furthermore, the EU used more coercive means to ensure stability through the presence of protectorate structures in Bosnia and Herzegovina (EUSR/OHR) and the interim administration in Kosovo (EULEX), which have been functioning with more or less success.

On the financial aspect, since 1991, the European Union has dedicated, through various assistance programmes, 6.8 billion euros to the Western Balkans. In 2000, aid to the region was provided through a new programme called Community Assistance for Reconstruction, Development and Stabilisation (CARDS). Through the programme 4.6 billion euros has been provided to this region in the period 2000 to 2006 for investment, institution building, and other measures to achieve four main objectives: (a) reconstruction, democratic stabilization, reconciliation and the return of refugees; (b) institutional and legislative development, including harmonization with European Union norms and approaches, to underpin democracy and the rule of law, human rights, civil society and the media, and the operation of a free market economy; (c) sustainable economic and social development, including structural reform; and (d) promotion of closer relations and regional cooperation among countries and between them, the EU and the candidate countries of central Europe (EU Council 2000).

In so doing, the European Union wanted to introduce reform processes in countries of the region to bring them closer to EU values, market and political environment, and to (re)build sustainable democratic systems which would in turn provide security and stability in the region, since the Western Balkan countries have been seen not only as strategic partners and eventually full Member States of the EU, but also as potential danger to the stability in Europe (and for the EU).

Nevertheless, the carrot-and-stick approach of EU conditionality towards Bosnia and Herzegovina is relatively well envisaged, a few major flaws have been detected, while rigid EU bureaucracy and lack of a single voice is not helping. Why is a relatively good approach to EU enlargement that proved acceptable in other cases not working well in Bosnia and Herzegovina?

Political Conditionality and EU Financial Assistance

In order to discuss the EU's approach towards Bosnia and Herzegovina in the wider Western Balkan context, we will use the overview of key elements of this process outlined by Džihić and Wieser (2011). Their principal observation was that

> [t]he political conditionality of the EU in Bosnia and Herzegovina had to respond to issues of challenged statehood and the dysfunctionality of state structures, challenges that had not been the subject of EU conditionality in earlier enlargement rounds. Although the state structure in itself has not been considered an obstacle to the EU membership of Bosnia and Herzegovina, it has been recognized that the limited functionality of Bosnia's central government is an obstacle to further progress towards accession … Consequently, EU conditionality is involved in issues of state building and building up "Bosnian stateness". (Džihić and Wieser 2011).[2]

---

2 Other elements relate to the following issues: 'EU conditionality and its implementation must be understood as an interactive process between the EU on the one hand, and the applicant countries on the other … We need to recognise that this interaction is limited to the elite level … The EU accession process is mainly limited to interaction between the political (or bureaucratic) elites of the EU and the political elites in the

As previously mentioned, the semi-protectorate structure in Bosnia and Herzegovina is headed by the Office of the High Representative of the international community in Bosnia and Herzegovina. The main task of the OHR since the Dayton Peace Agreement has been to ensure its implementation and interpret its provisions. The end of mandate of the High Representative (HR) and possible closure of the OHR are not very clear. The general understanding was that it would be closed down once the country was on an 'irreversible' road to the EU and NATO. But with the chances of 'irreversibility' getting slimmer after 2006, the international community started to run out of arguments for closing it down (Sarajlić-Maglajlić 2011: 56).

For some time now, the international community has been talking about the fazing-out and hands-off approach in Bosnia and Herzegovina. Some sort of *fatigue* is apparent; understandable, but also very dangerous. Like a single parent trying to discipline their child with empty threats (the stick) and overindulgence (undeserved carrot), without any other guiding authority to help develop social values, the child is most likely to become a spoiled brat, and exhibit behavioural problems, and become a problem for the community. This is exactly how ethno-nationalistic political elites in Bosnia and Herzegovina are beginning to behave.

The OHR is no longer the motor driving Bosnia forward, and it is too late for it to resume that role in any open-ended way (ICG 2009). The Peace Implementation Council (PIC) announced in 2006 that it wanted to close the OHR and rely henceforth on the EU. At that time, the period of progressive reforms was diminishing and Bosnia and Herzegovina was entering the 'magic circle of permanent crises' (Džihić 2010: 235). The PIC comprises 55 countries and agencies that support the peace process by assisting financially, providing troops for EUFOR, or directly running operations in Bosnia and Herzegovina. There are also a fluctuating number of observers. The PIC has a Steering Board which provides the High Representative with political guidance. In Sarajevo, the High Representative chairs weekly meetings between the Ambassadors to Bosnia and

applicant country. Thus, EU conditionality directs formal changes in the functioning of democratic institutions and structures, and thereby mainly relies on elected representatives to implement necessary reforms coherently. This inherently top-down process, at least in the Western Balkans, provides limited incentives to citizens to participate in the reform … By introducing the Copenhagen criteria the EU defined the framework of its political conditionality, but originally it remained rather broad and vague … Specifically in the Western Balkans the EU enlarged the scope of political conditions. In order to develop a strategy towards the countries in the region, which later became known as the Stabilisation and Association Process (SAP), a European Council decision in 1997 added further preconditions to the already well-established Copenhagen criteria … The principle of "own merits" (Council of the EU 2003) became the cornerstone of EU democratic conditionality by introducing country-specific SAP benchmarks and defining democracy-relevant conditions for each potential member country individually … The EU thereby became the main actor of external democracy promotion in the region, with overlapping processes of Europeanisation and democratisation' (Džihić and Wieser 2011).

Herzegovina and the Steering Board members. In addition, the Steering Board meets at the level of political directors three times a year. The Steering Board members are Canada, France, Germany, Italy, Japan, Russia, United Kingdom, United States, the Presidency of the European Union, the European Commission, and the Organization of the Islamic Conference (OIC), which is represented by Turkey. As observed elsewhere, '[t]he PIC hoped this would spur Bosnia to qualify faster for EU membership, but the opposite has happened: left largely to themselves, Bosnian leaders have become locked in a standstill, and some reforms have begun to unravel. Some argue that the shock therapy of an end to the OHR would have a salutary effect on politicians who have grown accustomed to being protected against the worst effects of their irresponsible behaviour' (ICG 2009).

Bosnia and Herzegovina is a country with newly introduced democratic mechanisms which have not come about as the result of internal democratic processes and people are not sufficiently prepared to be able to bear the sovereignty as should happen in a democratic society. Furthermore, dealing only with political elites (as often the international community and the EU have done) only suppresses democratization further. After Dayton, the international community and, more recently, the EU have headed political guidance, intended to raise awareness among people, democratic responsibility among political elites and the overall democratic culture in Bosnia and Herzegovina society. Well, it seems they are giving up – *The operation was successful but the patient died* – we tried reanimating with the *Carrot and Stick*, but it did not work!

Paddy Ashdown was the first HR to announce that he would be the last, right at the start of his mandate in 2002. His successor, Christian Schwarz-Schilling, conveyed the same message throughout his one-year mandate, while the next two High Representatives, Miroslav Lajčák and Valentin Inzko, were tasked with planning the transition from the OHR to the EU Special Representative (EUSR). The current Head of the EC Delegation to Bosnia and Herzegovina, Ambassador Peter Sorensen, presented his credentials to the Presidency of Bosnia and Herzegovina on 9 September 2011. Ambassador Sorensen is 'double hatted' as EUSR and Head of the EC Delegation to Bosnia and Herzegovina. No major improvement to the role of the international community in Bosnia and Herzegovina was made with Sorensen. Maybe the worst political crisis since the end of the war has been going on since the general election in 2010, and a long period is needed to make a coalition at state level and form a Council of Ministers of Bosnia and Herzegovina.

Belgium earned the award for longest period without government for 541 days of negotiations after the 2010 election, which beat war-torn Iraq (249 days without government) for the dubious world record. Well, even here, Bosnia and Herzegovina is not the best (as one joke goes: *You are so incompetent that in the competition for the 'most incompetent one' you would finish in second place – you cannot be first even there!*) but is in a strong second place with 480 days without government! Considering that Bosnia and Herzegovina is actually a (semi)protectorate, this story could be a strong candidate for a 'Believe it or Not' section. So far, the international community and the European Union have mainly negotiated reform processes and offered *carrots* and *sticks* to the political party leaders believed to have the most power to represent the respective constitutive peoples. We will later review the example of constitutional reform processes in Bosnia and Herzegovina and the role of the EU.

Even though it sometimes seems that the EU is not committed full time to Bosnia and Herzegovina, at least through regular Progress Reports the European Commission is showing quality analysis of the situation. Among the first conclusions of the 2011 Progress Report on Bosnia and Herzegovina was that overall, 'little progress was made by Bosnia and Herzegovina in improving the functionality and efficiency of all levels of government. One year after the general elections, a State-level Council of Ministers remains to be appointed. The political representatives lack a shared vision on the direction to be taken by the country. An effective coordination mechanism between the State, the Entities and Brčko District remains to be established as a matter of urgency regarding EU matters and the harmonisation of EU related legislation' (EC 2011: 10). The NGO Centre of Civil Initiatives' (CCI) interpretation of the 2011 EC Progress Report includes the following statements: 'Failure to reach political agreement and form the government at the state level hinders the country's progress on key reforms required for achieving further progress in EU integrations. The accession to EU requires functional institutions on all levels and effective mechanism for coordination of EU related issues' (CCI 2011). This demonstrates that the EU is aware that state institutions are weak and that they need them as strong partner for the enlargement process; however, it does not fully explain their approach in emphasizing negotiation and partnership with political party leaders and entities.

As a result of its specific historical, constitutional and evidence-based situation, collective rights in Bosnia and Herzegovina should be protected and taken into account. However, talks should not involve the political party leaders, but rather the democratically elected representatives in the legislative and executive branches, such as members of the Bosnia and Herzegovina Presidency (Bosniak, Croat, Serb – directly elected by the people) or delegates to the Bosnia and Herzegovina Parliamentary Assembly (House of Peoples and/or House or Representatives).

Firstly, people elect their representatives to protect their interests. Political party leaders are elected by their members (most political parties are run in totalitarian style with strong leader cults). On the other hand, even with lot of flaws, the Bosnia and Herzegovina Constitution does provide for some democratic mechanisms and institutions. For example, the House of Peoples of the Parliamentary Assembly of Bosnia and Herzegovina has 15 delegates, two-thirds of whom are from the Federation of Bosnia and Herzegovina (five Bosniak and five Croat delegates) and one-third comes from Republika Srpska (five Serb delegates). The nominated Bosniak and Croat delegates from the Federation of Bosnia and Herzegovina are elected by Bosniak and Croat delegates respectively in the House of Peoples of the Federation of Bosnia and Herzegovina, while the delegates from the Republika Srpska are elected by the National Assembly of the Republika Srpska. Even being discriminatory (see Sejdić and Finci *vs.* Bosnia and Herzegovina) this provides a platform to discuss collective rights and not to accept political party leaders as representatives of peoples.

But any such 'democratic mechanisms' become just a form. In practice, political party leaders of several of the strongest parties are placed above democratically-elected institutions and elected representatives since, by default, all elected representatives have to do whatever their party leader decides in the country. This situation is very much supported, and even encouraged, by the EU identifying these political party leaders as their partners in the reform process negotiation. Therefore, state institutions are continuously weakened and power is shifted to few ethno-nationalistic party leaders, suffocating democracy both in the political arena and within political parties.

This is how the European Commission assessed the progress of Bosnia and Herzegovina on the road to a European future and their success with the carrot and stick:

> Bosnia and Herzegovina has made limited progress in addressing the political criteria. In the areas of democracy and the rule of law, one year after the general elections held on 3 October 2010, the process of establishing executive and legislative authorities remains to be completed with the establishment of a State-level Government ... The lack of a credible process for the harmonisation of the Constitution with the European Convention on Human Rights remains an issue of serious concern ... The administrative capacity of the Parliament improved but lack of coordination between the State and Entity parliaments and political discord between the Entities has continued to hamper the work of the Parliamentary Assembly ... Bosnia and Herzegovina has achieved very limited progress in tackling corruption ... [P]olitical representatives in Republika Srpska frequently challenged the territorial integrity of the country ... As regards the economic criteria, Bosnia and Herzegovina has made little further progress towards a functioning market economy. (European Commission 2011)

Even after the Council of Ministers of Bosnia and Herzegovina was finally formed at the beginning of 2012, the parties in coalition are still not making any progress in political, economic or social reforms. Furthermore, current coalitions are falling apart because they have not been formed on the basis of a programme platform but rather on division of power and increase of their influence on public companies, agencies and institutions, including the Public Broadcasting System.

The ethno-political elites have discovered that the EU is no longer going to use the 'stick', while the 'carrot' alone is not working. The only form of punishment that the EU is exercising is to stall the progress of EU integration, thereby only punishing the people and further empowering the politicians to continue to keep their voters in vicious circle of fear that is just strengthening position of ethno-nationalistic political leaders. As seen, even with the 'carrot' there is an issue between the State and Entities on whether to accept it and how to use it.

Another serious issue, and probably the root problem, is corruption. Transparency International (2011) placed Bosnia and Herzegovina in 91st place on the corruption chart, with a score of 3.2. The lack of political will and inadequate anti-corruption efforts are hindering Bosnia's fight against corruption, says a joint report by the Center for Investigative Reporting (CIN) from Sarajevo and the Center for the Study of Democracy (CSD) from Sofia, a project funded by the EU. According to the report, levels of corruption in Bosnia started rising after 2000. New anti-corruption legislation and strategies were introduced at this time, and courts began investigating cases of corruption. However, convictions on charges of corruption are still rare, and high-level politicians and organized crime figures are notably absent from courtrooms. Even when measures are taken against such individuals, reversals of court decisions and release from custody are not unusual. As observed, 'the fight against corruption in Bosnia and Herzegovina consists of largely uncoordinated activities backed by no explicit local political will' (CSD 2012). Hence, it is very difficult to expect political elites to remove themselves from office and go to jail, since fighting corruption would apparently end up like that. This is another important argument to have in mind in understanding the current political landscape in Bosnia and Herzegovina, as it reveals how staying in power is more important than a future in the European Union for some political leaders.

Nevertheless, there were periods when everybody was thinking that better days were ahead, but, as always, something went wrong. Bassuener contends that at some point (2005–2006) the international community saw that Bosnia was moving irreversibly toward becoming a functioning state, and toward EU and NATO membership:

> It might move slower than we'd like to see, but it was only a matter of time until it got there. The international community could afford to continue dismantling the safety net that it had created for Bosnia. 2006 should have made clear this was not the case, but the international – especially EU – approach has remained on bureaucratic autopilot ever since, despite the deteriorating political situation. This has generated increasing uncertainty in Bosnia, as all the rules that previously applied after Dayton have been violated without consequences. The return of organized violence was unthinkable in early 2006. Four years later, it is no longer. There is a sense that the international community has no strategy for Bosnia, and is fixated on reducing its own responsibility. This projection of weakness has emboldened all those political leaders who espouse unfulfilled agendas to cross previously sacrosanct red lines. (Bassuener 2010)

Constitutional reform in Bosnia and Herzegovina has been the most challenging task for both local politicians and the international community. The international community understood that it was partly responsible for the very difficult constitutional framework which should have been a provisional transitional solution and not permanent tool for nationalists to keep people in fear and maintain the status quo. Furthermore, the EU saw at the beginning that it would be very difficult to have such a disorganized state as a Member State and requested this to be one of the conditions, but later it softened its position. A new constitution remains desperately needed to ensure equal human rights, freedoms, equality, rule of law and social and economic prosperity along with processes of truth and reconciliation in Bosnia and Herzegovina, to make acceptable consociation agreements and a functional State, and not be the ultimate playground and experiment for the international community.

In April 2010, the latest (serious) attempt by the EU and USA to help overcome the political crisis in Bosnia and Herzegovina failed, with key Bosnian political players rejecting US Deputy Secretary of State James Steinberg's and Spanish Foreign Minister Miguel Ángel Moratinos' efforts, during a visit to Sarajevo, to convince them to sign up to constitutional reforms and a reform agenda after the 2010 general election (Bieber 2010). Previously,

> the hasty and ill-prepared EU–US initiative at the Butmir NATO base next to Sarajevo airport in autumn 2009 contributed to creating the sense of crisis rather than alleviating it. International mediators had hoped to kick-start constitutional changes and secure a deal on the remaining criteria necessary for the Office of the High Representative in Bosnia and Herzegovina to be closed. While driven by desirable goals, the process was mishandled by the US and the EU. The initiative was designed to put pressure on the political parties to agree to major concessions, but most walked away from the talks and rejected the internationally proposed reforms when they realized that neither Sticks or Carrots were on the table, just a show of engagement. (Bieber 2010)

Following the aborted talks and a few further rounds of technical discussions, the so-called Butmir Process is dead. It was one in a line of significant failures of previous reform efforts, from the 'April Package' (of 2006, which was rejected in the Bosnia and Herzegovina Parliament) to the 'Prud Agreement' (a vague platform for constitutional reforms drafted in a small Bosnian village in 2008).

Despite several failed attempts to lead constitutional reforms in Bosnia and Herzegovina, the international community has even increased its pressure on Bosnia and Herzegovina's political elites after the Decision by European Court of Human Rights in Strasbourg in the case of Sejdić and Finci *vs.* Bosnia and Herzegovina. Bosnia and Herzegovina has been a member of the Council of Europe since 2002 and was therefore party to the European Convention on Human Rights. Furthermore, the Dayton Peace Accords incorporated rights and freedoms from the European Convention (along with 15 other international instruments on human rights) directly into the Bosnia and Herzegovina Constitution. The Bosnia and Herzegovina Constitution ensures direct applicability of the European Convention and has priority over other legislation in Bosnia and Herzegovina (Seizović 2008).

The case of Sejdić and Finci *vs.* Bosnia and Herzegovina (application No. 27996/06 and 34836/06) has increased pressure by the international community on Bosnia and Herzegovina's political elites to make the necessary constitutional reforms. In order to understand better the complexity of constitutional framework some explanation is needed. Both Dervo Sejdić (Roma national minority) and Jakob Finci (Jewish national minority) are citizens of Bosnia and Herzegovina. The Constitution of Bosnia and Herzegovina (Annex IV of Dayton Peace Accords), in its Preamble, makes a distinction between two categories of citizens: the so-called 'constituent peoples' (Bosniaks, Croats and Serbs) and 'Others' (Jews, Roma and other national minorities, together with those who do not declare affiliation with any ethnic group, including those who declare themselves as just Bosnians (and Herzegovinians)). The House of Peoples of the Parliamentary Assembly (the second chamber) and the Presidency are composed only of persons belonging to the three constituent peoples. Mr Jakob Finci enquired with the Central Election Commission of Bosnia and Herzegovina about his intentions to stand for election to the Presidency and the House of Peoples of the Parliamentary Assembly. On 3 January 2007 he received written confirmation from the Central Election Commission that he was ineligible to stand for such elections because of his Jewish origin.

In December 2009 the judgment of the European Court of Human Rights was published. The Court found that the applicants' ineligibility to stand for election to the House of Peoples violated Article 14 of the ECHR (ban of discrimination in the field of Convention rights) taken in conjunction with Article 3 of Protocol No. 1 (free elections), and that their ineligibility to stand for election to the Presidency violated Article 1 of Protocol No. 12 (general ban of discrimination). However, even two and a half years after the Decision of the Court, there have been no serious efforts by leading political parties to make the necessary constitutional reform and the international community is getting tired of offering carrots and showing sticks, since neither is really working any more. Why?

Conclusion

We have seen that the EU (alone or/and in partnership with others) is struggling to find a way (fast)forward for Bosnia and Herzegovina and to find the best use of the carrot and stick. Among other issues, international credibility took a big hit in 2007 when the EU signed a Stabilisation and Association Agreement with Bosnia and Herzegovina – a major step in the accession process – even though specific police legislation that the EU strenuously insisted on as a precondition had not been adopted (Bieber 2010). Furthermore, the High Representative was engaged in a related confrontation with the leader of the Republika Srpska. The closing of the OHR while two objectives (resolving the status of the Brčko District and dealing with state property) and one condition (full compliance with the Dayton agreement), identified by the PIC in 2008 as requirements, remain unmet would risk crippling the EU's ability to apply firm policies toward Bosnia long after the protectorate itself has ended. It would also weaken EU credibility throughout the region, notably in Kosovo (ICG 2009).

As Bieber (2010) suggests, the latest efforts with the 'Butmir Process' failed for three reasons: the EU and US were ill prepared, suggesting 'quick fix' solutions to the parties. Second, international mediators had little to offer in exchange for reform – just the 'carrot' on the paper – especially to those parties that would have lost out as a result. Third, by creating a sense of emergency in Butmir's army barracks, the mediators suggested that extra-institutional and quasi-coercive means might be used to change Bosnia's political structure. The EU and the US lacked the means and determination to back up such a high-risk strategy.

While the nationalist and uncooperative rhetoric of an ethno-nationalist political leader is not helping the process, the international community must take a large share of the responsibility. The EU, in particular, has taken a contradictory line. The criteria and conditions for further progress in the Stabilisation and Association Process have changed and remain unclear. The EU has been particularly ambivalent about constitutional reform, supporting (although not whole-heartedly) the first US-led efforts that failed in April 2006 and then subsequently stating that constitutional changes are not a requirement, but are necessary. Even if this might be right, it has spread damaging confusion (Bieber 2010).

Since citizens of Bosnia and Herzegovina see potential membership as *l'aboutissement du rêve* (the dream come true/solution for all problems) and are largely supportive of the process (VPI 2012), political elites often use EU accession in public discourse to please the voters. There can be no doubt that, although political elites declare their support of EU accession, no major political actor is willing to agree to yield powers for the sake of this alone, at least not without a serious carrot-and-stick game plan.

Furthermore, decisions on which level of government negotiates with the EU, and where standards are set and implemented, are not technical but profoundly political, and the EU continues to make the mistake of avoiding official institutions of the state. This is even more puzzling since the negotiations and partnerships with all other former Yugoslav republics were conducted only with state officials/institutions and not political party leaders or lower administrative units of the country. There are, of course, different constitutional frameworks in all countries, with the conditions in Bosnia and Herzegovina being very specific (historically, politically and constitutionally), but the principles are (or should be) the same for all on the road to the EU.

The EU's record of engagement so far has not been good: the failed police reform demonstrates that when the locus of power becomes more important than the substance of reform and when the requirements lack credibility, the EU is bound to fail. Consequently, effective EU engagement needs to focus not on one particular institutional set-up, but rather on clearly identifying what different institutional set-ups can (and cannot) engage with the EU during the accession process and once Bosnia becomes a Member State (Bieber 2010). Ever since the first elections after the war in 1996, there have been hopes that more moderate and cooperative parties would come to power, but more often than not, the opposite has been true – or, as was the case in the 2006 and 2010 elections, the 'moderates' turned out to be more uncompromising than the established nationalist parties (Bieber 2010).

It is crucial to emphasize that the effectiveness of the EU's conditionality approach depends on three factors: (a) the EU's readiness and level of dependability in pursuing conditionality principles; (b) the structural context of the State where conditionality is being applied (functionality of the state and its administrative capacities); and (c) the willingness and engagement of political elites consequently to undertake their obligations within the EU integration process (Džihić 2010: 232). All these levels are being seriously challenged in the case of Bosnia and Herzegovina.

To end on a more optimistic tone, the EU and the international community have done a great job in some segments, including stopping the war, providing security and stability in the regions, supporting the country's reconstruction and facilitating several successful state-building projects (such as regulation of the communications sector by establishing and supporting the Communications Regulatory Agency of Bosnia and Herzegovina, and defence reform), providing substantial financial assistance in reconstruction and reforms and given hope to the citizens of Bosnia and Herzegovina that a better future is possible with(in) the EU. Civil society is being supported and education for democracy has been promoted.

One successful example of the use of the carrot and stick is the Schengen zone visa liberalization process in which a set of reforms has been successfully implemented. This has demonstrated that when the goal is clear, relevant partners are identified along with who is leading the process and what the ultimate prize is – it all goes much easier. With the liberalization of the visa regime for the Schengen zone, young people have been given a sense of freedom and the ability to travel freely to study and learn. By exposing more citizens to other European countries Bosnia and Herzegovina will be able to have a better future with new educated generations and to adopt best practices and knowledge in various fields. Even though the process of brain-drain is still a concern, more cases of brain-gain are being reported. With the Dayton Peace Accords the international community has taken responsibility for the future of Bosnia and Herzegovina. The EU needs to stimulate the government and ethno-nationalist political elite to be more responsible and efficient by better use of the very powerful tool – the carrot and stick. The European Union does have both, it just has to find the best model of employment. The EU's approach towards Bosnia and Herzegovina should be simultaneously democratization, Europeanization and state building.

Nevertheless, I offer here a few principles for the future in line with Bieber's (2010) recommendations. First, constitutional amendments should only be introduced through the formal institutional process, not pushed through ad hoc meetings, to reduce the risk of spoiling tactics if only party leaders are included. Second, the EU should steer the process by providing advice and guidance, offering a clear menu of options to prepare Bosnia's institutional structure not only for accession negotiations but also for membership; it will not produce immediate results and Bosnia will continue to require EU patience and long-term engagement.

However, the main power remains in the people of Bosnia and Herzegovina who have managed to overcome great historical obstacles with a smile and good spirit still preserved. Slowly but surely, they are understanding political reality and the need to hold elected officials responsible for their promises and actions. After all, the citizens of Bosnia and Herzegovina have had only 20 years to play with democracy in very difficult circumstances. However, the European Union is the logical and natural partner for the present and future, and hopefully will stay fully committed to Bosnia and Herzegovina and the region.

References

Abazović, D., Mujić, A., Vehabović, F. and Vajzović, E. 2010. *Mjesto i uloga "Ostalih" u Ustavu Bosne i Hercegovine i budućim ustavnim rješenjima za Bosnu i Hercegovinu*. Sarajevo: Fakultet političkih nauka.

Aybet, G. and Bieber, F. 2011. From Dayton to Brussels: The Impact of EU and NATO Conditionality on State Building in Bosnia & Hercegovina. *Europe-Asia Studies*, 63(10), 1911–37.

Balfour, R. and Stratulat, C. 2011. *The Democratic Transformation of the Balkans*. EPC Issue Paper, No. 66. Brussels: European Policy Centre.

Bassuener, K. 2010. *Bosnia-Herzegovina, the EU and the International Community*. Available at: http://www.boell.eu/downloads/Bassuener_BiH_EU_Int.Community.pdf [accessed: 4 February 2012].

Bieber, F. 2010. *Constitutional Reform in Bosnia and Herzegovina: Preparing for EU Accession*. Available at: http://www.epc.eu/documents/uploads/1087_constitutional_reform_in_bosnia_and_herzegovina.pdf [accessed: 22 February 2012].

Bieber, F. and Keil, S. 2009. Power-Sharing Revisited: Lesson Learned in the Balkans? *Review of Central and East European Law*, 34(4), 337–60.

CCI 2011. *Council Of Ministers BiH January-September 2011: Report Summary*. Available at: http://www.cci.ba/files/user/docs/CAP%20II/Council%20Of%20Ministers%20BiH%20Jan-Sept%2011%20summary.pdf [accessed: 29 January 2012].

Coffey International Development 2009. *Assessment and Analysis of EU Accession Risk Factors for Bosnia and Herzegovina*. Available at: http://www.acips.ba/bos/uploads/dokumenti/Assessment%20and%20analysis%20of%20EU%20accession%20risk%20factors%20for%20BH%20final%20draft%20July%202009.pdf [accessed: 26 January 2012].

CSD 2012. *Countering Corruption in Bosnia and Herzegovina: 2001–2011*. Available at: http://www.csd.bg/artShow.php?id=16081 [accessed: 17 June 2012].

Domm, R. 2007. Europeanisations without Democratisation: A Critique of the International Community Peacebuilding Strategy in Bosnia and Herzegovina. *Journal of Southeast European and Black Sea Studies*, 7(1), 159–76.

Domm, R. 2011. Next Steps on Bosnia-Herzegovina: Key Elements to a Revised EU Strategy. *Journal of Southeast European and Black Sea Studies*, 11(1), 53–67.

Džihić, V. 2010. 'Evropa' u Bosni i Hercegovini – Bosna i Hercegovina u 'Evropi' – EU-kondicionalitet i interna previranja u Bosni i Hercegovini: Gdje i zašto smo stali odnosno kako da se pokrenemo?, in *Bosna i Hercegovina petnaest godina nakon Daytona*, edited by D. Abazović and S. Hammer. Sarajevo: Fakultet političkih nauka, 231–44.

Džihić, V. and Wieser, A. 2011. Incentives for Democratisation? Effects of EU Conditionality on Democracy in Bosnia & Hercegovina. *Europe-Asia Studies*, 63(10), 1803–25.

EU Council 2000. Council Regulation (EC) No. 2666/2000 of 5 December 2000 on assistance for Albania, Bosnia and Herzegovina, Croatia, the Federal Republic of Yugoslavia and the Former Yugoslav Republic of Macedonia, repealing Regulation (EC) No. 1628/96 and amending Regulations (EEC) No. 3906/89 and (EEC) No. 1360/90 and Decisions 97/256/EC and 1999/311/EC.

European Commission 2011. *Bosnia and Herzegovina 2011 Progress Report*. Brussels, 12 October 2011. SEC(2011)1206.

European Parliament 2009. *Resolution on Srebrenica*. Available at: http://www.europarl.europa.eu/sides/getDoc.do?pubRef=-//EP//TEXT+TA+P6-TA-2009–0028+0+DOC+XML+V0//EN&language=EN [accessed: 1 December 2011].

Freyburg, T. and Richter, S. 2010. National Identity Matters: The Limited Impact of EU Political Conditionality in the Western Balkans. *Journal of European Public Policy*, 17(2), 263–81.

Hoare, A. 1997. The Croatian Project to Partition Bosnia-Hercegovina, 1990–1994. *East European Quarterly*, 31(1), 121–39.

Holbrooke, R. 1998. *To End a War*. New York, NY: Random House.

Ibrahimagić, O. 2010a. Ustavna rješenja u Bosni i Hercegovini i "problem" sa federalizmom, in *Bosna i Hercegovina petnaest godina nakon Daytona*, edited by D. Abazović and S. Hammer. Sarajevo: Fakultet političkih nauka, 13–18.

Ibrahimagić, O. 2010b. *Politički sistem Bosne i Hercegovine*. Tom I. Sarajevo: Promocult.

ICG 2009. *Europe Report No. 198*. Available at: http://www.crisisgroup.org/en/regions/europe/balkans/bosnia-herzegovina/198-bosnias-incomplete-transition-between-dayton-and-europe.aspx [accessed: 17 January 2012].

Kasapović, M. 2005, Bosna i Hercegovina: Konsocijacionalna ili liberalna demokracija? *Politička misao*, XLII(5), 3–30.

Kent, S.A. 1999. The History of Bosnia-Herzegovina. *Faces*, 15(8).

Kunrath, B. 2010. Od zaštitne mjere do prepreke za dalje djelovanje: Entitetsko glasanje u Bosni i Hercegovini, in *Bosna i Hercegovina petnaest godina nakon*

Daytona, edited by D. Abazović and S. Hammer. Sarajevo: Fakultet političkih nauka, 19–47.

Lavić, S. 2012. Bosanskohercegovačka dijaspora i migracije, in *Refugees and IDPs: Scientific and Research Challenges and Perspectives*, edited by E. Vajzović. Sarajevo: Fakultet poliričkih nauka, 65–80.

Malcom, N. 2011. *Bosna: Kratka povijest*. Sarajevo: Buybook.

Misita, N. 2007. *Osnovi prava Evropske unije*. Sarajevo: Pravni fakultet Univerziteta u Sarajevu.

Mujkić, A. 2008. O perspektivama postdejtonske Bosne i Hercegovine, in *Ustavne promjene u Bosni i Hercegovini*, edited by D. Jerinić. Banja Luka: Nezavisne Novine, 131–56.

Pejanović, M. 2010. Protivrječnosti odlučivanja u parlamentarnoj skupštini Bosne i Hercegovina: Problem uspostave konsenzusa unutar parlamentarne većine, in *Bosna i Hercegovina petnaest godina nakon Daytona*, edited by D. Abazović and S.Hammer. Sarajevo: Fakultet političkih nauka, 77–87.

Petrović, J. 2012. Izbor prebivališta kao faktor socijalne integracije raseljenih osoba, in *Refugees and IDPs: Scientific and Research Challenges and Perspectives*, edited by E. Vajzović. Sarajevo: Fakultet poliričkih nauka, 53–63.

Rupnik, J. 2011. The Balkans as a European Question, in *The Western Balkans and the EU: 'The Hour of Europe'*, edited by J. Rupnik. Paris: EU Institute for Security Studies, 17–30.

Sarajlić-Maglić, D. 2011. Bosnia and Herzegovina after the Elections: A Tale of Disillusioned Optimism, in *The Western Balkans and the EU: 'The Hour of Europe'*, edited by J. Rupnik. Paris: EU Institute for Security Studies, 47–58.

Seizović, Z. 2008. Konstitutivni narodi i ustavne promjene, in *Ustavne promjene u Bosnia and Herzegovina*, edited by D. Jerinić. Banja Luka: Nezavisne Novine, 157–76.

Sejdić and Finci *vs*. Bosnia and Herzegovina (application No. 27996/06 and 34836/06). Available at: http://cmiskp.echr.coe.int/tkp197/view.asp?action=html&documentId=860268&portal=hbkm&source=externalbydocnumber&table=F69A27FD8FB86142BF01C1166DEA398649 [accessed: 15 February 2012].

Smajić, M. 2011. *Medjunarodne organizacije i sigurnost Bosne i Hercegovine u postdejtonskom periodu*. Sarajevo: Dobra knjiga.

Stahl, B. 2011, Perverted Conditionality: The Stabilisation and Association Agreement between the European Union and Serbia. *European Foreign Affairs Review*, 16(4), 465–87.

Transparency International 2011. *The Corruption Perceptions Index Ranks*. Available at: http://cpi.transparency.org/cpi2011/results/ [accessed: 15 June 2012].

US Department of State 2012. *Background Note: Bosnia and Herzegovina*. Available at: http://www.state.gov/r/pa/ei/bgn/2868.htm [accessed: 5 February 2012].

Vajzović, E. 2004. Smrt ili novo radjanje nacija. *Novi pogledi*, III(I), 18–21.

Vajzović, H. 2008. *Jezik i nacionalni identiteti*. Sarajevo: Fakultet političkih nauka.

VPI 2012. *Javno mijenje BiH o procesu europskih integracija 2009–2012*. Sarajevo: Vanjskopolitička inicijativa BH. Available at: http://www.vpi.ba/bos/sadrzaj/dokumenti/Javno_mijenje_BiH_o_procesu_europskih_integracija_2009-2012.pdf [accessed: 16 June 2012].

Žarin, I. 2007. EU Regional Approach to the Western Balkans: The Human Security Dimension. Available at: http://www.doiserbia.nb.rs/img/doi/0025-8555/2007/0025-85550704513Z.pdf [accessed: 4 February 2012].

Chapter 8
Democratization in Kosovo: The Role of International Institutions

Leandrit I. Mehmeti

At the Thessaloniki Summit of June 2003, the European Union (EU) made it clear that the future of the Western Balkans is closely tied up with Europe and that regional integration is the key to the future of the region. International institutions have played a crucial role in political development in the Balkans, from restoring peace to institution-building support and assistance. Organizations such as the UN, NATO and the EU were mandated to re-establish security and assist in institution building and economic revival. Since the declaration of independence by Kosovo authorities in 2008, the UN has been phasing out and the EU has taken a leading role in monitoring the progress of Kosovo institutions towards their democratic development. Kosovo has undergone great development in institution building, albeit that the process has had deficiencies and challenges, especially after the war.

This chapter analyses the democratization development in Kosovo over the last two decades. It explores the political and ideological developments during the 1990s in the form of political movements and parties and their development after the war of 1999. In doing this it analyses the role of international institutions in the process and explores and attempts to assess the development of today's Kosovo institutions. Later, this chapter looks at the EU's role in the process and argues that it is crucial not only to the democratic development of institutions but also to regional stability and economic development and sustainability. It argues that the EU integration process has been the main impetus behind democratization in the region, and in Kosovo, and that any lessening of such a drive from the EU itself due to 'enlargement fatigue' could be very dangerous for the region. This chapter concludes that the EU holds the key to peace in the region by continuing its support of democratization and economic development.

Introduction

The European Union considers the Balkan region of particular importance when it comes to EU integration for many reasons, the main one being the political turmoil and conflicts throughout the last century and especially during the 1990s and the potential that still exists for such conflicts to regenerate. A completely integrated Europe, as a stable political and economic entity in a global perspective, will face great challenges with regard to its functionality and political seriousness if the Balkan region is left out of the EU and becomes unstable again. On the other hand, if the Balkan region is integrated into the EU, the conflicts and wars are less likely to regenerate and national borders are also less likely to play a significant role.

Nevertheless Kosovo–Serbia relations remain one of the greatest regional challenges of the EU, and a formula for accepting two countries, one of which refuses to recognize the other, into a community of countries (the majority of which have recognized that country),[1] is complicated, tense and very challenging. The EU is making every effort to find the solution to this challenge, as reflected in the latest technical talks between the Kosovo and Serbian governments on normalizing the relationship between them. The EU is the most important actor in such an enterprise. However, because of the historical relationship of the Balkans with Europe, the EU is often viewed with scepticism from many political actors in the Balkans.

This chapter elaborates on these issues by looking at the current political and security challenges and obstacles. It illustrates the involvement of international actors in the process, to continue in more depth on the democratization process and development in Kosovo. In doing this, the chapter will be structured in three parts. The first part will analyse briefly the installation of international institutions in Kosovo following the NATO intervention against Yugoslavia/Serbia in 1999, offering a brief historical background. The second part will analyse the democratization process in Kosovo, looking at its beginning in the Socialist Federal Republic of Yugoslavia (SFRY) following the collapse of communism, the changes of the political system in 1989 and the abolition by the Milošević regime of the autonomy that Kosovo enjoyed after the 1974 constitutional changes in SFRY. The review will include the establishment of political movements and parties to continue the development of such political groupings and institutions after the war. The third part will analyse the challenges of Kosovo institutions regarding EU integration and its relationship with Serbia.

Installation of International Presence in Kosovo

Historical Background

Peter Novick argues about objectivity when writing history, drawing a line between the 'knower and known, fact and value, history and fiction', adding that truth and patterns in history are 'found' and not 'made', and that the historian should refrain from taking a judgmental role that could 'degenerate into that of advocate or even, worse, propagandist' (Novick 1988: 2). Writing about the history of the Balkans

---

1 Kosovo has been recognized by 85 countries around the world as of December 2011 and by 22 EU Member States. EU members that have not yet recognized Kosovo as an independent state are Cyprus, Greece, Romania, Slovakia and Spain.

is very challenging. The historical evidence provided by many historians in the Balkans varies quite a lot and is often contradictory. In short, the history written in the Balkans is often hard to distinguish from folklore.

When considering the role of different individuals and institutions, it is apparent that the conflicts in the former Yugoslavia were political rather than ethnic (Judah 2000, Kuperman 2008, Malcolm 1998, Mertus 1999, Vickers 1994). Malcolm (1998: xxvii–xxvix) elaborates on the role of Slobodan Milošević and other political personalities in triggering those wars for the personal and political gains of their respective republics, or factions within them, using particular political platforms (such as different memorandums of the Serbian Academy of Sciences and Arts, in 1937 and 1986),[2] defining these conflicts as more political than ethnic. Mertus points out that these ethnic groups have lived in harmony and shared a mutual respect for their rights and cultures for centuries and have even fought together against other powers, such as the late Ottoman Empire's occupation of the Balkans, suggesting that there is still a chance that they can live together in peace and harmony, although it will be difficult in the aftermath of the conflicts (Mertus 1999: 5–9). Historically, conflicts between peoples of the Balkans started to appear after the first signs of weakness in the Ottoman Empire, with its involvement in wars against Russia in the late eighteenth and nineteenth century. The arbitrary decisions of the then European powers on the borders of the Balkans in the late nineteenth century proved that the establishment of peace in the region was fragile, and conflicts reappeared as recently as the last one that followed the disintegration of the SFRY. The states that were created and established with the blessing of the European powers in 1878 seem to have had repercussions on the events of 1912, the events in 1945, and again in the more recent events of 1990–99. These historical antecedents tend to be ignored by the EU itself and other international actors when projecting different decisions for the region and while developing policies. As Mayer suggests, 'understanding the antecedents of the problems facing the Balkans can help statesmen avoid past mistakes' (Mayer 2005: 238).

NATO Intervention and the UN Installation

For eight years (1990–98) Kosovo maintained a particular kind of peace that was built between the Albanian political projection for an independent Kosovo 'that would be accomplished through the divine intervention of the West, and especially of the USA, and the Serbian reality of ruling Kosovo violently' as Shala (2003: 13–15) would observe. The passive peaceful resistance against Serbian rule preached through Kosovo Albanian leader Ibrahim Rugova started to change

2 For more information on the content and role that both documents, and especially the Memorandum of 1986, played in the establishment of the nationalist policy of the Serbian authorities in the late 1980s and early 1990s, see Judah (2000), Malcolm (1998) and Mertus (1999).

after the Dayton Peace Accords, where the Kosovo issue was not mentioned or included in the agreement. Different actors emerged within Albanian society. In the beginning they supported policy changes, initially called 'active peaceful resistance', later to be modified to supporting different approaches of active resistance, including guerrilla operations against the Serbian police and army.

The Kosovo Liberation Army (KLA) that emerged as an organization spread throughout Kosovo within days. There are no serious statistics about the number of members, as the KLA did not establish a vertical line of organization. What was important, however, was that this organization enjoyed the support and respect of many of the Albanian people in Kosovo and it played an important role in support of a more active involvement of the international community in the Kosovo crisis.

After the escalation of the conflict and the displacement of hundreds of thousands of people internally in Kosovo, and later outside of it, a humanitarian catastrophe was imminent. The Milošević–Holbrooke agreement, reached in October 1998, mandated an International Verification Mission under the Organization for Security and Cooperation in Europe (OSCE), whose mission was to verify the ceasefire between the Yugoslav Army and the KLA. The ceasefire was not respected and scenes of massacres and displacement of the civilian population pushed the international community to take more drastic measures towards reaching a peace agreement between the conflicting parties. They 'locked in' the parties in conflict in the Rambouillet Castle near Paris to negotiate until they could reach an agreement that would stop the conflict and would permit a political solution in the near future. Although the Albanians were now on the verge of fighting for independence, eventually an agreement was reached that called for an international military presence and political arrangements to end the conflict. The Rambouillet Accord would put Kosovo into the position it had enjoyed with the SFRY Constitution of 1974, with some differences in foreign relations, taxation policy, education and health policy, and most importantly it recognized that this was an accord for a period of three years, after which the Kosovo people would express their will on a political solution through a referendum. Apparently, it was so extremely difficult to insert such a paragraph from the Albanian side, and especially from the KLA representatives, that the conference was about to be called a total failure.[3] The agreement was signed only by the Albanian delegation. The refusal of the Serbian delegation to sign the agreement resulted in the NATO-led air strikes against Yugoslavia to end the Kosovo conflict.

3 Rambouillet Agreement, Chapter 8, Article 1, point 3 stated: 'Three years after the entry into force of this Agreement, an international meeting shall be convened to determine a mechanism for a final settlement for Kosovo, on the basis of the will of the people, opinions of relevant authorities, each Party's efforts regarding the implementation of this Agreement, and the Helsinki Final Act, and to undertake a comprehensive assessment of the implementation of this Agreement and to consider proposals by any Party for additional measures.'

The Kosovo case marked the first time that a substantive part of the international community went beyond its own principles and intervened against a sovereign state without the authorization of the UN Security Council. Intervention ended the war and brought peace, yet it left the political situation unresolved. Russia's threat to veto the proposed supervised independence for Kosovo[4] took the issue out of the Security Council for the second time. The International Court of Justice (ICJ) responded to Serbia's question, deposited through the UN General Assembly, with a ruling that the declaration of the independence of Kosovo did not violate international law (ICJ 2010). Will this lead to a different approach to the international principles in the future? One cannot be certain at this point, yet a revision of these principles seems to be happening.

However this may be seen and interpreted, one thing is certain, the intervention ended the century-long oppression by the different Yugoslav/Serbian regimes of the Albanian population living as a minority in Kosovo. It ended the systematic violation of human rights that happened there especially during the 1990s, and it ended the war in June 1999 and opened the path for the installation of the UN civilian mission (UNMIK) to help with the administration and rebuilding of Kosovo. More than 800,000 refugees returned to Kosovo within weeks. UNSC Resolution 1244 legalized the international military and civil presence. Despite some wording changes, Resolution 1244 restated major conditions stipulated in the Rambouillet Agreement, except for the three-year period clause (Klasnja 2007: 15) and some changes with regard to the military presence of Yugoslavia/Serbia in Kosovo, with their numbers varying from a few thousands, according to the Rambouillet Agreement, to full withdrawal of all security forces, according to UNSC 1244.

UNSC Resolution 1244

UNSC Resolution 1244 was approved in June 1999 following previous UNSC resolutions (1160, 1199, 1203 and 1239) on the Kosovo issue that were approved in the period from March 1998 until June 1999. It authorized an international

---

4 Negotiations for the final political status of Kosovo began in 2006 under Martti Ahtisari, former President of Finland, between Kosovo and Serbia authorities. Ahtisari had been called upon to mediate and facilitate the negotiations by the UN Secretary General Kofi Annan, following an earlier report for the progress of UNMIK's standards from Kai Aide, a Norwegian diplomat. Aide suggested that the implementation of the policy of 'standards before the status' put in place by UNMIK in 2002/3 would be impossible to achieve before the status was clearer and Kosovo authorities were given wider competencies. The negotiations for the final status lasted for about two years and Ahtisari proposed a supervised independence with special rights and links for the Serbian minority towards the authorities in Belgrade. Ahtisari's proposal had been accepted by the Albanians after a lot of pressure as a compromise for full independence but was rejected by the Serbian authorities as it was considered a precedent and an annexation of a part of their territory.

civilian and security presence. The security presence (under NATO command) would ensure the security of the borders of Kosovo, the complete withdrawal of the Serbian/Yugoslav police, army and paramilitary troops and ensure that these forces would not return to Kosovo unless as specified by the resolution and would supervise a complete demilitarization of the KLA.[5] The civilian presence would have the competencies to promote and establish substantial autonomy and self-governance in Kosovo.

The territorial integrity of the FYR (Federal Yugoslav Republic) and the Rambouillet Accord principles are other important aspects of this resolution that have been discussed thoroughly by different international, Kosovar and Serbian legal experts, including during the hearings in the ICJ sessions (in November–December 2009) on the legality of the declaration of independence by the Kosovar authorities. The resolution recognizes the territorial integrity of the FYR, including Kosovo. However, it calls for a final political settlement, recognizing the temporariness of the resolution and the principles of the Rambouillet Accords and its determination to act for these purposes under Chapter VII of the Charter of the United Nations.[6]

Different actors from all sides have interpreted Resolution 1244 in different ways. The Serbian interpretation emphasizes the points about the territorial integrity of FYR/Serbia over Kosovo and 'the return of the Serbian and Yugoslav personnel', which is usually interpreted as the return of the army, police and administration, despite the specifications of the resolution including the numbers (hundreds, not thousands) that were to be supervised by the international security presence. Interpretations from Kosovo tend to focus on the points concerning the temporariness of the resolution and the call for a final political settlement according to the Rambouillet Accord, which was signed by the Kosovo side in 1999. UNMIK's interpretation concentrated on the establishment of democratic institutions in Kosovo and the evaluation of these developments. Yet, as it happened, according to political developments during the UNMIK administration in Kosovo, those evaluations were not the sole responsibility of the UNMIK but rather of other international actors involved in the Kosovo issue such as different individual countries.

---

5 Article 6 of the Annex 2 of the resolution states: 'After withdrawal, an agreed number of Yugoslav and Serbian personnel will be permitted to return to perform the following functions: Liaison with the international civil mission and the international security presence; Marking/clearing minefields; Maintaining a presence at Serb patrimonial sites; [and] Maintaining a presence at key border crossings' (UN 1999).

6 'Determined to ensure the safety and security of international personnel and the implementation by all concerned of their responsibilities under the present resolution, and acting for these purposes under Chapter VII of the Charter of the United Nations.'

Political Developments since 1989

The Democratization Process in the SFRY

The collapse of communism in the SFRY was followed by development of tense political situations that led to military conflicts involving different ethnicities. Military conflicts have been quite common in the post-communist world and, as Horowitz (2003: 25) observed, within 28 successor states eight have been affected by large-scale military conflicts. As a multi-ethnic state with fragile ethnic problems, hidden underneath Tito's established governance with a 'brotherhood and unity' slogan, the SFRY went through a different democratization development compared to other Eastern European countries: a radicalization of different ethnicities. Yugoslavia went from being the most open and liberal society in the region, the socialist country with the highest per capita income in the communist world, and the country deemed most likely to join the European Community, to being behind the regional curve and a site of growing political conflict which led to violent warfare (Gagnon 2010: 23). Despite being labelled as ethnic wars, these conflicts and wars were more the result of struggles for economic and political reform. These struggles started in Kosovo as early as in the 1960s (to continue with different intensity during the 1970s and 1980s) and then spread to Croatia in 1971, bringing about the first signs of the fragility of the Yugoslavian project. In Kosovo the main demands centred around equal rights in the federation, the consideration of Albanians as a nation and the consideration of Kosovo as a federative unit, while in Croatia they centred around demands for more autonomy for the republic. Both these struggles were crushed with violence that resulted in many political prisoners.

Albanians in Yugoslavia were the third nation in number, after Serbs and Croats; yet they did not enjoy the same nation status as the other republics.[7] Although Kosovo (and Vojvodina) became federative units with the constitutional changes in 1974, they still remained provinces of Serbia and were denied republic status, therefore the status of nation. The theoretical answer to why Kosovo and Vojvodina were denied republic status was that republics were entities of the nations as opposed to nationalities (terminology coming from the Soviet Union doctrine). A nation could form a republic in a federation and therefore have the right to secession. The practical answer was political: a Kosovo republic could

7 Albanians in Yugoslavia were scattered in three republics: Serbia (Kosovo), Macedonia and Montenegro. In Kosovo they made up 80–90 per cent of the population of nearly 2 million. In Macedonia they made up 20 per cent of the population of 2 million. A small percentage also lived in Montenegro and Serbia proper. For more details see Critchley (1993: 441). The percentage of Albanians living in Macedonia varies according to statistics. According to official results of the census in 1994, Albanians constituted 22.9 per cent of the total population. However, the Albanians claim that they constituted 40–50 per cent of the population (Babuna 2000: 80).

secede from Yugoslavia and join Albania. Zoran Pajić describes this situation as creating two categories in SFRY, the hosts and historical guests. The hosts (nations) were Serbs, Croats, Slovenes, Montenegrins and Macedonians, while the historical guests (national minorities), were all other ethnic groups living in the SFRY (Pajić 1995: 162).

Federalism in the late 1980s in Yugoslavia was faced with three crucial and interrelated factors that would seal the fate of the SFRY: identity politics, the relationship between the federal understanding of further centralization or decentralization of political powers between the federation and the federative units, and economic disparities between federative units.[8] Skalnik Leff argues that all communist multinational states shared commonalities on how identity politics shaped the opening up of the political systems with regard to freer public discussion and organization and to competitive elections (Skalnik Leff 1999: 208–10). Further, she questions whether ethnofederalism truly qualified as federalism, if federalism is generally understood as an institutional arrangement where authority and functional competences are shared among different levels of government. This is quite a reasonable question starting from the fact that the centre accorded official recognition to the identity of the dominant titular nationalities in each republic, which, in many cases, had unpleasant consequences for ethnic mobilization (Skalnik Leff 1999: 209). The general idea of the establishment of a particular 'Yugoslav' identity for all the nationalities of SFRY simply proved to be unsuccessful. Fink-Hafner (1995: 344–5) argues that the problem of maintaining the federal state was also directly related to the failure to create one society within the framework of the former Yugoslavia.

The attempts to create a unified Yugoslav identity, both national and political, from the early period following the First World War, were seen through different lenses by different nationalities, which ultimately led to the creation of two different visions of federalism: one that supported a more centralized federal structure that would give more power to the federal level and less to the republics (favoured by Serbs) and the other that supported a more decentralized federal structure (favoured by a majority of nations in the SFRY that saw the federation as being led by Serbs) (Critchley 1993: 442). The constitution of 1974 resolved

8 Economic disparities in the SFRY were quite evident between the republics and autonomous provinces, with Slovenia having the highest general income per capita and Kosovo the lowest. Political development and stability, human rights issues, authoritarian discriminatory governance towards different ethnic groups and the level of democratization of the institutions and the geographic positioning of some republics contributed to economic development in different parts of Yugoslavia. As this chapter will not include a thorough analysis of the economic aspect of differences within the SFRY, for more details see Bebler (2002), Critchley (1993), Fink-Hafner (1995). On the theoretical aspects of how frustrated national ideals, political institutions and conflict and war can distract governments and institutions from a direction of economic and market reform agendas, as was the case with SFRY, see Horowitz (2003).

the tensions between these two visions of federalism, for a period, creating a more decentralized structure and allocating more power to the republics and to the autonomous provinces of Kosovo and Vojvodina. The new constitution acknowledged that the SFRY comprised eight federative units. However, as the situation developed, dissatisfaction increased, especially among Serbs, who viewed themselves as losers, and considered all the changes as conspiracy against them, especially when Kosovo and Vojvodina were carved out of Serbia to become autonomous regions and federative units. Such changes became the main points to be used by Milošević in the late 1980s for nationalist mobilization among Serbs, firstly in Kosovo, and then in other parts of Yugoslavia where Serbs were living as a minority. Such developments ultimately led to the collapse of federalism and the opening up towards political pluralism in the SFRY.

Yugoslavia had developed subsystems of grass-roots self-government and self-management with different levels of development and engagement in different republics (Bebler 2002: 130). This development was more evident in Slovenia and in Croatia and less so in the other republics and autonomous units. Elements of pluralism existed in different professional bodies, such as the academy of sciences, universities, and around some literary and social-science journals. In Kosovo, the League of the Kosovo Writers and the Institute of Albanian Studies (the latter affiliated with the University of Priština) tended to be the more vocal instruments in the call for more decentralization of power in Kosovo and the upgrade of its political status to a republic and political pluralism. In 1989, as in other parts of Yugoslavia, different groups, affiliated with those groups and institutions that demanded and argued for different political changes, started to appear and call themselves political movements or parties. In January 1990, the League of Communists of Yugoslavia (LCY) in Belgrade publicly acknowledged that the single-party system in Yugoslavia was dead (Bebler 2002: 131).

Nationalist parties were victorious in elections in 1990 in Yugoslavia across the six republics and in Kosovo. The elections created a particular situation that triggered the radicalization of ethnicities in the SFRY, usually though the exclusion of minorities from access to government or the boycott by minorities of national government institutions (Birnir 2007: 13). In other cases in the SFRY the election led to ethnicization of the political environment: the collapse of multi-ethnic coalitions with radical candidates winning the elections led to what Rabushka and Shepsle called the 'bankruptcy of moderation' (cited in Kubo 2009: 2–3). Similar situations of minority boycott were created in most of the federative units. What happened in Croatia, Bosnia and Herzegovina and Serbia (Kosovo) was a general boycott of government institutions by minorities (namely Serb in Croatia and Bosnia, and Albanian in Kosovo). Minorities often organized themselves in structures parallel to the government, in non-violent ways of protest against the newly elected government or, as has happened, in violent protest up to a full-scale war against the government. Birnir (2007: 13) argues that the main problem for maintenance of moderate inter-ethnic relations was not the ethnicization per se, but the exclusion of ethnic minorities from the government, which may result in radicalization in some cases but not always. This seems to have happened at the beginning of pluralism and democratization in Kosovo in the late 1980s and beginning of the 1990s. With the constitutional changes and Milošević's suppression of the autonomy that Kosovo had enjoyed since 1974, the Albanian minority was not only excluded from the governance institutions but from all public (state) institutions including health care and education.

Democratization in Kosovo

Kosovo had a different transition compared to other former communist countries after the fall of communism in the region that influenced the development of pluralism and democratization of the political party. This influence determined Kosovo's different path for many reasons, but the main ones can be attributed to changes in the political, social and economic environment in Kosovo after the constitutional changes imposed by the Milošević regime.

The initial demands of most of the political groups were a repetition of the previous ones for republic status for Kosovo within SFRY, yet they quickly shifted towards the struggle for full independence following similar developments in Slovenia and Croatia. All parties supported the establishment of a democratic system with free elections and the protection of minorities' rights. In the 'shadow' elections in 1990 (that were considered illegal by the government in power in Kosovo) the party led by Ibrahim Rugova, of the Democratic League of Kosovo (LDK) won the absolute majority of the votes. The government that was established operated mainly from abroad, soon to be known as the government in exile. Rugova's government was limited to some level of control and organization of the education and health care sector, funded on a voluntarily basis through a symbolic tax collected in Kosovo diasporas.[9]

The nature of the 'shadow' government in Kosovo and its very existence under Milošević's rule made it appear more as a fiction than an efficient institutional

9 All staff of the former education system worked for nearly 10 years without pay with the only objective being to keep the education system up and running and enable students to continue learning, as schools in Albanian language and with the curriculum that Kosovo enjoyed before the autonomy was abolished, were totally banned and closed to students of Albanian background from Milošević's regime. Health care was in the same situation and was still operating under great limitations. The government in exile introduced a 3 per cent annual income tax per capita, a voluntary tax on all Kosovo Albanian individuals on working contracts all around the world to support their activities, which were limited to education and health, and the other activities of the government. Eventually, Kosovo Albanians responded massively and signed up for support for an indefinite period of time. It is not clear how many people signed up for such an enterprise, but unofficially the speculation circles around 700,000 to 1.3 million. It is also not clear as to the spending of such a budget, as the administration of this revenue was not transparent. The education and health system seemed to continue to work without pay. Universities operated with the revenue they generated from tuition fees, paid in full by students.

government to foster democratic transformation and development. Despite the limitations that the government had while operating in Kosovo, the fact that political and social pluralism started in Kosovo in the form of political parties and movements as well as some not-for-profit organizations were signs of a democratization process. The tremendous growth of the LDK and the absolute majority that was ensured in the first elections also created a particular situation with regard to political ideological development and the internal political party democratization, not only within the LDK but also within other political organizations.

Development of Political Concepts

The new political groups of Kosovo Albanians in Kosovo adjusted their policies according to the new political environment and circumstances, and basically developed two dominant political concepts and discourses with regard to the future of Kosovo. The main political group, LDK, established a political concept that would project the struggle for independence from Serbia through peaceful resistance. Such a concept meant the general boycott of state institutions and the internationalization of the Kosovo issue. Further, this internationalization meant a large-scale campaign to inform the international community of the political and human rights situation created in Kosovo after the changes and to convince it to support the right to self-determination of the Albanian people of Kosovo in their struggle for independence. This concept was seen as the only way to ensure the continuation of such a struggle and, above all, to prevent an escalation of the conflict that was thought would have grave consequences for the people of Kosovo.

The alternative concept (rooted back in the 1920s) saw the political future of Kosovo as reunited with Albania rather than as an independent country, and sought freedom from Serbian rule through active resistance (supported by guerrilla operations against the Serbian Army and police) rather than peaceful resistance, imagining a possible direct involvement of Albania in such an enterprise. Such a concept saw the involvement of the international community in the Kosovo issue, imagining Albania as the most important actor. As realistic as this might be considered, developments during the 1990s proved an important point in this respect. Albania, after all, was not the country some of the leaders of such movements, who had been educated and fed with Enver Hoxha's (Communist leader, dictator of Albania) propaganda, imagined, and was not going to get itself involved actively in the Kosovo solution. Therefore, this concept did not enjoy the support of the majority of people in Kosovo.

As the situation developed in the SFRY and especially in Serbia, the political discourse established by the LDK was considered as the most realistic one, despite the fact that it did not pose any serious threat to Milošević's rule. Milošević's group could continue its policies, as situations developing with the conflicts in Croatia and Bosnia were far more important for the international community than

Kosovo. The created status quo enabled the 'Western leaders to diplomatically ignore Kosovo', as Blumi (2002) would put it. Whether the LDK political discourse might have prevented the conflict in the early 1990s is still a matter of further discussion and analysis.

LDK versus Other Political Groups in Kosovo

The establishment of the political concept of 'passive peaceful resistance' by the LDK created a particular situation with regard to democratization. The democratization process had been limited to only two elements: the internal democratization of political parties and the media. Other forms of democratization of institutions were simply impossible in Kosovo due to the lack of institutions, because of the general boycott and non-recognition of Milošević's institutions in Kosovo by Kosovo's Albanian political establishment.

It is difficult to observe any serious development of political ideologies in the political parties in Kosovo during the 1990s. It may be suggested here that political parties in Kosovo during that time remained more as political movements than parties, and their political ideological development was very limited due to the common political aim and objective, that of the ultimate independence for Kosovo. Having a common political objective to establish a state seems to have inhibited the development of ideologies about how such a state would function and develop and what kinds of policies the political parties would follow with respect to other elements of the state that would ensure a better quality of life for the people.

The political concept of the LDK dominated the political environment especially in the first years of its establishment until the Dayton Peace Accord for Bosnia. The growth of the LDK marginalized all political parties in Kosovo and basically they were only orbiting around the LDK's development. This created a particular situation with the LDK seeing itself as the strongest and the most developed party in Kosovo thus halting its own internal democratization process, which was also reflected in the internal development of other political parties. The party conventions, for example, would be organized in the same way all the time, with the same leadership being legitimized and not opening opportunities for new faces. This also created a particular political culture within the political environment in Kosovo, that of the leader of the party, who was seen as the person to drive the reforms and the democratization development of the party. The development of the party systems was therefore more actor-driven than structure-driven. In this sense, any initiatives for further policy debates or party institutional development and reform were doomed to end in a cul-de-sac if they were not supported by the leader. Such a phenomenon could be attributed to the long tradition of authoritarian institutions functioning in a former communist country or to the particular political situation in the 1990s in Kosovo which meant that the struggle for independence from Serbian rule was far more important than internal party development.

If such a justification can be valid, then this is as naïve as it may be from the simple fact that after the war in 1999 Serbian rule over Kosovo ended and the political party development was still continuing in the same tradition, despite the fact that the assistance of international institutions such as the OSCE was never greater than in that period. Such a situation created a difficult position for those reformists, albeit symbolic in numbers, who wished to oppose such internal party developments and push for more open, transparent and democratic ways of functioning of the party structures. This political culture was also reflected in other parties during that period and after the war. The death of Ibrahim Rugova in January 2006 proved how much political parties in Kosovo had lost with the lack of internal political party development, which is indicative of the level of democratization. The main political parties are still bound in a particular way to their leaders. After Rugova's death, the LDK was split into three groups, with more perhaps to happen in the near future. Similar situations can be observed for other parties as well, bearing in mind the difficulties they face while operating in the absence of their leader, as in the example of the Aleanca për Ardhmërinë e Kosovës [The Alliance for the Future of Kosovo] (AAK), the party of the former Kosovo Prime Minister, Ramush Haradinaj, indicted by the International Criminal Tribunal for former Yugoslavia for war crimes. Should an internal political party democratization process have been fostered during the period before the war, the party (the LKD) perhaps would not have gone through such changes.

Time after time, other political groups/parties would appear and disappear. These parties could not escape marginalization by the LDK and disappearance from the scene was quite common. The explanation for this phenomenon could be both social and political. Social, because society in general has viewed the LDK's political discourse as the most viable to prevent conflicts (that were already under way in Croatia and Bosnia) and to win the support of the international community through the policy of internationalization. Political, because the leadership of the LDK felt secure in the Kosovo political environment with no real opposition in sight and any initiative of democratization would ultimately lead to the formation of different political concepts and mechanisms, and therefore would possibly weaken the LDK's political security.

Such developments were witnessed in Kosovo after the Dayton Agreement, which seriously weakened the LDK's position, not so much through internal political party reforms as through development of different political circumstances, especially when an agreement on education between Rugova and Milošević, sponsored by Sant'Egidio Community in Rome, on the use of educational facilities by Albanian students was reached but never implemented. Political opponents saw this as a total failure of Rugova's policy and as a dangerous indication that the future negotiations with Serbia, which were promoted by Rugova's LDK, were shifting from political international actors and institutions to humanitarian non-governmental organizations. This shift strengthened the more radical political groups in Kosovo because society started to observe possible cracks in the independence project.

The main mechanism of the LDK in maintaining the popularity of its political concepts during the 1990s was through the media. In the early 1990s, the media in Kosovo were influenced by the LDK and would promote only the LDK's view. The appearance, in 1997, of *Koha Ditore*, a daily newspaper funded with the support of the Soros Foundation, brought a different point of view from the other daily newspaper that was a clear voice of Rugova's policy, appearing more independent and offering more objective analysis about different political factions and social trends. This change with regard to the media had crucial consequences for the political and social developments to follow in Kosovo. This media outlet challenged the LDK's view and reported the position of Kosovo internationally, breaking the conviction that had been created through the LDK's media that the situation in Kosovo would be resolved through the LDK's policies. It did this by reporting that Kosovo was not on the international community's agenda and that the chances were that if it did get on it, it might not be for negotiation of independence but rather for a political settlement that would be limited to some sort of autonomy within Serbia. This daily newspaper is still considered one of the most independent media outlets in Albanian.

The Democratization Process After the War

Robert Keohane (1998: 82) points out that analysing world politics in the 1990s is to discuss international institutions, and this is very valid for the Balkans of the last two decades. The international civilian presence, in the form of UNMIK, was given the task of rebuilding Kosovo from scratch. Institutions were completely non-functional and Kosovo Albanians had failed to fill the institutional vacuum created following the withdrawal of Serbian security forces and administration in the way they had agreed while at the Rambouillet Conference. It is still not clear whether the LDK's refusal to participate in the Provisional Government of Kosovo (PGK, or the Unity Government as it was called) with Hashim Thaqi (KLA political director and the head of the Albanian delegation in Rambouillet) as Prime Minister or the lack of an invitation from the KLA to do so resulted in the LDK's exclusion from the government or whether the phenomenon of the 'leadership' played any role in such a situation or any other agreement with other 'actors' involved. This issue is more covered with speculation rather than by any serious evidence.

UNMIK claimed all administrative, executive, judicial and legislative powers in Kosovo, mandated from UNSC Resolution 1244 to establish democratic institutions for self-government. UNMIK considered the Provisional Government of Kosovo (PGK) as 'illegal'. However, the general mood of the population and the acceptance of the international presence, especially the NATO forces, in Kosovo was very positive. The international community had a very good advantage regarding institution building in Kosovo, considering the difficulty of building functional institutions from scratch in a society that had not participated in any institutional form for a decade and that lacked qualitative education even in the former Yugoslavia regarding institution building.[10] The positive mood towards the international community's presence was advantageous in the sense that they could launch a great educational programme of democratization of the institutions in Kosovo, if they had the will to do so. Observing the developments in Kosovo in this regard, one comes to mixed conclusions. This first political clash between the Kosovo Albanian political establishment and the international community concerned the latter's neutrality towards Kosovo's political status. The project of independence still remained the main political objective of most of Kosovo's political groups and the parties experienced a similar situation as before the war with regard to their development.

The International Community, the Democratization Process and Institution Building

To ensure smooth operation of the mission UN Secretary-General Kofi Annan oversaw the setting-up of the organizational structure of UNMIK through a four-pillar structure: UNMIK, to deal with all administrative issues in Kosovo; UNHCR, to coordinate the return of refugees and humanitarian assistance; the EU, to deal with economic reconstruction; and the OSCE, entrusted with the task of institution building aimed at 'strengthening the capacity of local and central institutions and civil society organisations … promoting democracy, good governance [and] organising elections' (Narten 2009: 136). The OSCE's task comprised elements such as training and raising 'awareness and involvement of citizens in social and political change in Kosovo by strengthening the development of local citizens, … professional, cultural and other associations, … programs to facilitate conditions that support pluralistic political party structures, political diversity and a healthy democratic climate, … training of government officials and executive and administrative officers in procedures of good governance' (Narten 2009). Since the Yugoslav crisis in the 1990s, the EU's involvement in the region had been limited more to humanitarian aid and crisis management, and the handling of the crisis had been quite disastrous, as Papadimitriou (2002: 186) would note. Yet, the involvement in the aftermath of the Kosovo crisis, being an integral part of the UNMIK structure, was a sign of a shift of management and involvement in the region, and EU institutions were a special help to the UNMIK (Weiss and Kalbacher 2008: 331).

The UN presence in Kosovo was unique in the sense that it claimed absolutely all powers of a state. The democratically elected institutions were able to exercise only limited responsibilities and the real powers remained with the Special Representative of the Secretary General (SRSG), who enjoyed unlimited executive

10 The University of Priština established its department of Political Sciences and Public Administration only after the war. After the changes in the constitution of 1974 in the SFRY, Kosovo was given the opportunity to open the university for the first time with particular departments in humanities and natural sciences

and legislative powers. The UN presence in Kosovo understood institution building in Kosovo literally as their exclusive experiment by not taking seriously the political groups in Kosovo, the social and political circumstances, the recent history of the 1990s and the post-conflict environment. Yet, despite the limited inclusion of political groups and actors in the institutional life of Kosovo, they still had the opportunity to undergo some real development. The main trouble that comes with such a situation is the level of measurement one can apply when analysing the democratization development without full inclusion of the political actors in institutional life. When something is evaluated in negative terms in security or governance matters, which institutions are to be held accountable, UNMIK or the Kosovo institutions?

The presence of international institutions in Kosovo, and especially the OSCE, contributed to the start of democratization development in political parties, local civil administration, civil society, the local judiciary, and the police force and media led by the OSCE with the assistance of a number of specialized international organizations. Training seminars that were organized and structured for all these levels, especially on political parties and public administration, included education on democratic practices and principles and multi-ethnic cooperation and reconciliation. They tried to cover all the structures of political parties and public administration. Bearing in mind the post-war environment in Kosovo, the challenges that these international organizations and institutions faced were quite serious, especially in multi-ethnic cooperation and reconciliation.

The challenges on this issue have been twofold. On one hand was the ethnicization of Kosovo's polity in general, although the Constitutional Framework, adopted during UNMIK's governance, and the Kosovo Constitution after the declaration of independence, allocated reserved seats for minorities in the Kosovo Parliament[11] and local councils followed the examples of many EU countries of the policy of positive discrimination. The Serb minority, especially the hard-line political actors in northern Mitrovica, boycotted and did not participate (and still do not) in the Parliament and other legislative bodies. Some Serb representatives from other Serbian enclaves in Kosovo are part of the government heading some ministries. On the other hand, Belgrade exercised, and still does, a very heavy pressure in the Serb minority in Kosovo, encouraging them not to participate in institutional life in Kosovo and to boycott cooperation with UNMIK and now with the EU-led Rule of Law mission (EULEX). As Obradović observes, the problem of the virtual separate development is forced from above, particularly from Belgrade and Kosovo Serbian authorities, as a continuing protest gesture against Kosovo independence. Based on NGOs' reports from the region, Belgrade is preventing the integration of Kosovo Serbs in Kosovo institutions, creating situations where more and more young people leave to study or work in Serbia proper because of

11 The Kosovo Parliament has in total 120 seats, 20 of which are allocated for minorities, with 10 seats for the Serb minority. They can also get more seats if they get a sufficient number of votes during the election process.

the difficulties that such pressures create (Obradović 2009: 2). The pressures are of a different kind and include the withdrawal of financial benefits up to the point of direct threats and intimidation through its security services operating clandestinely in Kosovo.[12] Despite the symbolic Serb participation in governmental institutions, political parties in Kosovo still remain established along ethnic lines with few multi-ethnic parties or party programmes in place.

The same is observable at the NGO level. Although more than 2,000 local NGOs were registered in Kosovo after the war, few of them are engaged in any multi-ethnic activities, and only a few of them can be considered as serious NGOs who still continue to operate despite the limitations on international financial aid and grants. This suggests that many NGOs that were operational in Kosovo while international financial assistance was available during the period of the aftermath of the conflict were established more to participate in the financial benefits rather than in purely civic matters. However, those that are still active represent a considerable membership of a number of people and serve as observers towards both the Kosovo polity and the international presence. Some of them even investigate corruption charges concerning different officials in the government at all levels as well as about different officials of the UNMIK and other international institutions, and lately of the EULEX. Their attempts on these matters, however, are still limited to different awareness-raising campaigns against these phenomena, and some cases with very high profile political leaders have ended in the courts after EULEX was pressured to lay charges for corruption against some individuals. Although it is hard to suggest that such charges were a result of awareness-raising campaigns from some local NGOs, no doubt they contributed to a better climate for accountability.

The important factor in these issues with civic organizations is their ability to raise the awareness in the general public challenging both international institutions and the government. The most prominent of such social movements (turned into political movement) is Vetëvendosja (VV). This movement has been able to mobilize a great number of young people in Kosovo and are a great opposition voice for many controversial policies. Albin Kurti, its leader, a former leader of students in the protest of 1998 before the war, has been charged and arrested

---

12 The Serb minority in Kosovo (those living in the northern part of Kosovo and other enclaves throughout the territory) enjoy financial benefits in the structure of different allowances, pensions and salaries for those working in local councils, health and education institutions in the enclaves. They have been and still are also being paid by the Government of Kosovo for their services. However, Belgrade is threatening to cut their benefits if they continue to take salaries from the Government of Kosovo, as there are some voices, especially in the enclaves, to participate and start their full integration in the institutions of Kosovo. There was a great shift in the last local elections held in November 2009, where a great number of Serbs participated in electing their representatives in some local councils where they make up the majority in enclaves in central and southern Kosovo. They have been condemned by Belgrade and other Serb political parties in northern Kosovo.

many times by the police. In one of the protests that this movement organized, the UNMIK police (Romanian police) killed two participants when it violently crushed the protest.[13] Arrests and charges for destruction of public property and other minor offences against this movement have certainly contributed negatively to the image of UNMIK, EULEX and the Government of Kosovo, especially bearing in mind the current general mood of much of the population, especially young people, about these institutions before and after the declaration of independence. For example, voter participation in post-war Kosovo was very high initially but has declined drastically over the years. In local elections in 2000 the turnout ranged from 79.9 per cent to more than 90 per cent (OSCE 2000) in different municipalities. In the 2001 assembly elections the turnout dropped to 64.3 per cent (OSCE 2001). In 2004 and 2007 the turnout was 53.57 per cent and 40.10 per cent (OSCE 2004, 2007), to increase at the last elections in 2010 to 64 per cent (KQZ 2010). The decline before 2010 can be seen as an expression of dissatisfaction and lack of belief in the political parties and in the UNMIK structures by the majority of the population, who view them as institutions that did not deliver what the people expected. Corruption allegations, the stagnation of economic development, the high rate of unemployment, the chaos created in public services, nepotism in the institutions, are all elements that have caused the decline in voter participation (before 2010) and trust in the UNMIK to fight corruption, especially when corruption charges circulated about some UNMIK members as well.

The increase in voter participation in 2010 could be seen as a result of the institutional crisis that called for the extraordinary elections, following a decision by the Constitutional Court that the President of Kosovo, Fatmir Sejdiu, had breached the Constitution by holding two public offices (of the president of the country and the political party). Another interpretation of this increase may be that these were the first assembly elections in post-independent Kosovo and the political environment was experiencing the emergence of new political groups such as Vetëvendosja. Voter participation increased with more young people participating in the elections.

The opposition on the other side, having been in the government before and having faced corruption charges, still does not look to be prepared to seriously challenge the government, although one can observe a debate culture in the parliament. The only constructed and the most voiced opposition in today's Parliament of Kosovo is VV, with a quite well-established politically profiled programme. This movement differs from other small political groups that support

---

13 This movement, for example, attacked especially the UNMIK regulation 2000/47 – On the Status, Privileges and Immunities of KFOR and UNMIK and Their Personnel in Kosovo, where local courts have absolutely no jurisdiction over the international staff of UNMIK and KFOR, whose personnel enjoys a general immunity. A number of cases of rape or murder by international staff working in UNMIK or KFOR and corruption charges involving millions of euros have gone unpunished.

only the reunification of Kosovo with Albania as the ultimate final solution to the Kosovo question, in the sense that it supports Kosovo independence yet asks for the right to reunification with any state that the Kosovo people would express through a referendum.

Good Governance in Kosovo and Organized Crime

Despite the fact that political parties received substantial and extensive education programmes on democratic principles and good governance from OSCE and other international organizations, it is quite difficult to observe any changes in their performance and efforts in establishing institutions based on good governance practices. Different elements and factors related to corruption and organized crime make good governance difficult to apply in Kosovo.

The transfer of power from one party or coalition to another in Kosovo as a result of elections has generally happened in a smooth and calm manner and there have not been any tensions between the parties, regardless of their political views and lines. This has demonstrated a relatively good culture of power transfer and acceptance of the election results. Although this could be attributed to OSCE monitoring and organization of elections in Kosovo since the end of the conflict, the last local elections in 2009 were organized by the Kosovo Government and the smooth operation and acceptance of results and transfer of power have been evaluated highly and praised by international organizations and the EU. However, the last parliamentary elections in 2010 were not as smooth as the local elections and were followed with numerous irregularities and allegations of fraud (KIPRED 2011: 5). Despite such allegations, the elections were repeated in the disputed municipalities and the results were accepted and certified.

Power transfers in Kosovo may offer a good image of democratization development, but public accountability and good governance practices suggest the opposite: political actors tend to understand their power more in terms of personal gain than as a public responsibility to ensure a better living standard for the population and to put in place stable institutions to cope with the political challenges of EU integration. The many corruption allegations involving particular government officials, including some high-ranking ministers, suggest that the political elite in Kosovo is slipping through the fast tracks of accumulating personal wealth through corruption and other criminal methods. These issues are seriously endangering the image of Kosovo and are further creating a very tense political situation inside the country, contributing greatly towards general public dissatisfaction.

Organized crime in the Balkan region is a very serious matter, especially for the EU, as it is on Europe's frontier and is a crossroads of different criminal activities connecting east and west, along with other countries of Eastern Europe. Balkan organized crime groups function in very complex networks involving many relevant actors, such as state security institutions like intelligence organizations, military and police, political leaders, paramilitary groups, religious leaders, and business leaders in state-owned companies. As Clark notes, Balkan organized groups conduct nearly the same operations that one would see anywhere in the world, thus providing goods and services that the governments of those societies are unable to provide or may limit or restrict (Clark cited in Stanchev 2004: 242).

During the 1990s, organized crime groups flourished throughout the Balkans and especially in the former Yugoslavia during the conflicts. Embargoes and sanctions necessitated the smuggling of different goods, such as fuels and weapons, and human trafficking of people looking for ways out of the conflict zones. Warlords usually presented themselves as uncompromising nationalists defending each nation's cause but at the same time established great cooperation networks with other warlords in support of each other, regardless of ethnic or political background and worked together when it was in their interests (Jović 2009: 18). Glenny describes in more details the cooperation between different groups of the criminal underworld in the Balkans and especially of those groups that already had the capacities to use different state security services (Glenny 2008).

However, the region is extremely stigmatized for its role in drugs, human or weapons trafficking or other forms of organized crime. There are other reports that suggest, on the contrary, that the Balkan region is one of the safest in Europe with regard to the levels of crime against people or property. The UN Office of Drugs and Crime (UNODC) suggests that the number of conventional criminal activities is falling in every Balkan country and that this trend is likely to continue as the countries are strengthening their justice systems, raising their living standards and becoming more integrated with the rest of Europe. The UNODC attributes these developments and improvements in dealing with organized crime to greater regional stability and democracy and closer integration with the EU through the EU's mechanisms (UNODC 2008: 5).

In Kosovo, organized crime is spread through all those networks mentioned earlier. However, Kosovo enjoys a different environment with the EU-led Rule of Law mission, EULEX, present in Kosovo. The EULEX seems to be the only force for the moment able to fight corruption and organized crime, as it would be very difficult for any new political power that might emerge in Kosovo to do so. EULEX authorities have started investigations of corruption allegations against a significant number of officials in local and central levels of government, including former officials from previous governments. Although such allegations date back to 2010, EULEX has still not completed any of the investigations. This seems to be a crucial time for the EULEX to assist institutions in fighting corruption and organized crime. A failure to continue with these investigations and bring those responsible to justice would put the EULEX into a very difficult position and the public will start to doubt its power, competence and will to tackle the real problems of governance in Kosovo.

After Independence

The role of international institutions in Kosovo has been crucial to restoring peace and assisting in the institutional building process and democratization after the war. The presence of these institutions has ensured that Kosovo did not slip into post-conflict anarchy but upheld law and order, for all its isolated problems; although a big contribution in this sense can be attributed to the population itself and their tradition of hospitality and mutual respect, as Abdela observed (2003: 209).

The work of these institutions became complicated as they were deployed in an environment where a majority of people fought and made sacrifices for independence with a posture of not recognizing that independence. This created a particular situation with regard to security matters as well. UNMIK maintained the status quo with regard to the progress of the political objective of the main Kosovo Albanian political parties, especially with the policy of 'standards before status', a policy that was doomed to fail before it even started. Once the main political objectives are considered to be endangered, and if ethnic tensions are still present, the situation degenerates to violence rather than dialogue. The riots of March 2004 are an example of such an eruption of violence that resulted in loss of lives on both sides and destruction of Serbian property and religious buildings. Such events have also shifted the approach of the international community towards the final political status of Kosovo. Kai Aide, a Norwegian diplomat, recommended to UNMIK a reassessment of the policy of 'standards before status'. Aide recommended that such policy was impossible to implement before the status became clearer and Kosovo authorities were given wider competencies. This opened the way to status negotiations facilitated by Martti Ahtisari, former Finnish President, that resulted in conditional independence for Kosovo.

Kosovo has entered a new phase of democratization and political development. Kosovo's independence has been declared and recognized by a significant number of countries, including most of the Western countries. The independence objective has taken a different path for the political parties that aspire to head the country, that of the consolidation of such independence by securing more recognition. Yet there also seems to have been a further development by political parties and different concepts with regard to perspectives on the integration of Kosovo in international institutions such as the EU or NATO. The Parliament of Kosovo has been enriched with new political groups that have brought new political concepts and have created a more vocal and more colourful opposition. There are also signs that some parties are refreshing their membership with new faces. Civil society has been filtered quite a lot with some serious NGOs and research institutions remaining that observe and monitor all institutions in Kosovo, both national and international. Relations between civil society and institutions are also entering a phase where public debate is very common and often controversial. NGOs in Kosovo seem to be quite vocal in exposing corruption affairs. Media development seems to still be as it used to be ten years ago, with most of them affiliated to political parties, yet there are still some media outlets that tend to remain independent and report on sensitive issues with regard to governance.

Kosovo institutions are also in a different position regarding the decision-making process after independence, although such decisions are often influenced by international institutions and other particular international actors. International institutions' competencies are limited to recommendations rather than arbitration of political decisions. The best example to illustrate this is the decision of the Kosovo Government to impose reciprocity measures with regard to trade against Serbia and Bosnia and Herzegovina. As a signatory of the Central European Trade Agreement (CEFTA) in 2006,[14] Kosovo was facing trade blockades from Serbia and Bosnia and Herzegovina after its declaration of independence as neither of these countries recognize its independence. In July 2011, the Kosovo government decided to apply reciprocity measures after the failure of Belgrade to respect the agreement reached in Brussels days before to recognize Kosovo's customs stamps. To implement this decision the Kosovo Government sent its police special forces to take control of two of its border crossings in northern Kosovo that had been taken under control by Serbs on 19 February 2008, following Kosovo's declaration of independence from Serbia (on 18 February). The Government of Serbia openly supported the actions of Serbs in Kosovo. The International Crisis Group reported on those events:

> On 19 February, a well-organised group, almost certainly including Serbian interior ministry (MUP) personnel and armed with guns and plastic explosives, was transported in buses and other vehicles to attack in succession the Brnjak and Jarinje border and customs posts in north Kosovo, drove away UNMIK and Kosovo police (KPS) personnel and blew up and set fire to buildings, equipment and police cars. The Serb mayor of Zubin Potok and the deputy mayor of Leposavic accompanied the assailants. That evening Minister Samardzic declared the attacks "not pretty", but in line with Serbian government policy. (ICG 2008: 8)

The Kosovo Government's decision to recover these border crossings triggered a great dissatisfaction in the Serbian community that resulted in road blockades to reject such measures.

Recent events in northern Kosovo have brought to the surface many political problems related to security and EU integration perspectives for both Kosovo and Serbia with regard to fulfilling the Copenhagen criteria. These events also illustrated the political positions of both governments. The EU is playing a crucial role in Kosovo–Serbia relations, especially after the events in northern Kosovo. The success of the EU at these talks remains to be seen. Agreements on some technical matters have been reached, such as on the movement of people and the

---

14 This agreement was signed in 2006 by UNMIK on behalf of Kosovo (CEFTA 2006).

recognition of personal documentation, and even a sort of compromise with regard to regional cooperation and presentation, with Koscvo being represented with a footnote (indicating UNSC Resolution 1244 and the ICJ ruling of the declaration of independence of Kosovo) alongside its name. This compromise has been interpreted as a victory in both Kosovo and Serbia for their domestic publics by the governments because of its content. The oppositions in both countries interpret this differently. In Kosovo, the government is blamed for purchasing a ticket to EU candidacy status for Serbia while not bargaining for a better carrot in return from the EU. Opposition in Serbia is accusing the government of entering on the path to recognize the independence of Kosovo. No matter what the interpretation, it is still not clear whether this compromise will contribute to better relations between the two countries. This compromise will possibly continue to be interpreted and used in different ways and there were occasions when both delegations walked out of different regional meetings during March 2012 because of complaints that the footnote was not properly read. The EU here is the main actor to settle such disputes by being clearer in its approach towards the Balkans and pressurizing the countries by using its 'civilian powers' (Whitman 2002) or its 'magnetic force' (Rosecrance 1998) to influence the developments it wants to see in the Balkans.

The integration of the Serbian community in institutional life in Kosovo is crucial to its democratization development. Yet this is not something that is entirely in the hands of national or international institutions in Kosovo. Without the support of the Serbian Government this seems to be quite impossible. This is an element where the EU, using its enlargement policy and its 'transformative powers' (De Ridder, Schrijvers and Vos 2010: 239), can play a crucial role. The approach of the EU suggests that it is on the right path, yet the difficulties that the EU itself faces when it comes to harmonization of foreign policy can endanger such an approach.

Conclusion

International actors and institutions have played a crucial role in the political development in the Balkans. The region has been a crossroads of many political interests of world powers and their influence is still crucial and evident with regard to the construction of national interests of many Balkan countries. Considering a rationalist approach in the enlargement context, states would want to join the EU as long as the political, economic and security interests were more beneficial than the expected costs. On the other hand, an international organization/structure such as the EU would admit new members as long as they brought economic and political benefits, and if this was not the case then they would admit new members as long as the risks of leaving them out were greater than the cost of admitting them (De Ridder, Schrijvers and Vos 2010: 240) As for the EU, they could discern security and economic reasons and the EU's influence in world affairs (De Ridder, Schrijvers and Vos 2010). The conflict in the former Yugoslavia proved the fragility of the security. Economic mismanagement, political instability or social unrest could hamper the transformation, and a population disillusioned by harsh reforms could choose to go back to authoritarian and nationalist regimes, creating a threat to the European Union's stability and security (De Ridder, Schrijvers and Vos 2010).

The Serbian political elite tends to ignore the fact that there was a war and a structured ethnic cleansing campaign in Kosovo, and that there was a situation of apartheid in the 1990s. The Serbian political elite seems to have been caught in the web between the possible will for political reform and its past involvement in all the conflicts in the former Yugoslavia, with what has been established as a national policy based on a socially constructed reality that was implanted in the Serbian public about the 'divine' right to defend Serbs in other countries. A possible shift out of such a political approach is difficult, yet not impossible. It has happened with many countries after conflicts.

The political elite in Kosovo, on the other hand, also tends to ignore the fact that, despite the massive exodus of the Serbs that fled together with Serbian security forces in June 1999 following NATO deployment, those who remained were subject to a campaign of humiliation and revenge from many Albanians who returned to Kosovo. A post-conflict society recovery requires that all political elites look towards the future and acknowledge the consequences of the conflict, and not ignore them but try to work on ameliorating them.

With regard to Kosovo and Serbia, the main international actors that both these countries are tied to when it comes to defending their political objectives seem to be not the EU but the USA and Russia. The reasons for this are quite clear when one takes a look at the recent history of the Balkans. In this respect, once the national interest might be considered at risk, both Serbia and Kosovo will look towards other international actors for support and not the EU.

In practical terms, the EU is, nevertheless, the ultimate address where the countries of the Balkans want to settle, as there seems to be a general agreement among political elites that the EU can serve as a guarantee of the stability of the region. The regeneration of the Balkan conflict seems unlikely if it is to be between parts of a supranational structure such as the EU. The EU integration policy triggered the democratization transformation and reform in all successor countries of the former Yugoslavia, albeit still in its early phase. Considering a social constructivist approach, where material costs and benefits are not enough to explain the eastern enlargement of the EU (Fierke and Wiener 1999, Sedelmeier 2000, Sjursen 2002), but non-material concepts such as common identity, values, norms and ideas are more important, then the enlargement of the EU towards the Balkans should happen faster and should be a 'question of when not if', as Dorian Jano (2008: 143) puts it. Challenges are great, bearing in mind the political situation between Kosovo and Serbia, yet a continuous pressure on both sides for a moderate and creative solution is necessary. Nevertheless, this pressure is only possible and effective when the political elites in both countries begin to take the situation more seriously. Without clear signals for the prospect of integration into the EU, these political elites will hardly make progress into further stages of cooperation, mutual agreement and reconciliation.

References

Abdela, L. 2003. Kosovo: Missed Opportunities, Lessons for the Future. *Development in Practice*, 13(2–3), 208–16.

Babuna, A. 2000, The Albanians of Kosovo and Macedonia: Ethnic Identity Superseding Religion. *Nationalities Papers*, 28(1), 67–92.

Bebler, A. 2002. Slovenia's Smooth Transition. *Journal of Democracy*, 13(1), 127–40.

Birnir, J.K. 2007. *Ethnicity and Electoral Politics*. Cambridge: Cambridge University Press.

Blumi, I. 2002. A Story of Mitigated Ambitions: Kosovo's Torturous Path to Its Postwar Future. *Turkish Journal of International Relations*, 1(4). Available at: http://www.alternativesjournal.net/volume1/number4/blumi.htm [accessed: 28 November 2011].

CEFTA 2006. *Agreement on Amendment of and Accession to the Central European Trade Agreement*. Available at: http://www.cefta2006.com/cefta-agreement [accessed: 20 December 2011].

Critchley, W.H. 1993. The Failure of Federalism in Yugoslavia. *International Journal*, 48(3), 434–47.

Fierke, K. and Wiener, A. 1999. Constructing Institutional Interests: EU and NATO Enlargement. *Journal of European Public Policy*, 6(5), 721–42.

Fink-Hafner, D. 1995. The Disintegration of Yugoslavia. *Canadian Slavonic Papers*, 37(3–4), 339–56.

Gagnon, Jr. V.P. 2010. Yugoslavia in 1989 and after. *Nationalities Papers*, 38(1), 23–39.

Glenny, M. 2008. *McMafia: Crime Without Frontiers*. London: The Bodley Head.

Horowitz, S. 2003. War After Communism: Effects on Political and Economic Reform in the Former Soviet Union and Yugoslavia. *Journal of Peace Research*, 40(1), 25–48.

ICG 2008. Kosovo's First Month. *Update Europe Briefing No. 47.*

ICJ 2010. *Accordance with International Law of the Unilateral Declaration of Independence in Respect of Kosovo*. 22 July 2010.

Jano, D. 2008. EU–Western Balkans Relations: The Many EU Approaches. *The Journal of the International University Institute of European Studies (IUIES)*, 2(1), 143–60.

Jović, D. 2009. Croatia after Tudjman: The ICTY and Issues of Transnational Justice, in *War Crimes, Conditionality and EU Integration in the Western Balkans*, edited by J. Batt and J. Obradović-Wochnik. Paris: EU Institute for Security Studies, 13–27.

Judah, T. 2000. *Kosovo: War and Revenge*. New Haven, CT: Yale University Press.

Keohane, R.O. 1998. International Institutions: Can interdependence work? *Foreign Policy*, Issue 110 (Spring), 82–96.

KIPRED 2011. *Kosovo National Elections 2010: Overview and Trends*. Priština: KIPRED.

Klasnja, M. 2007. The EU and Kosovo: Time to Rethink the Enlargement and Integration Policy? *Problems of Post-Communism*, 54(4), 15–32.

KQZ 2010. *Central Election Commission: Report 2010*. Available at: http://www.kqz-ks.org/SKQZ-WEB/al/zgjedhjetekosoves/zgjedhjetepergjitshme.html [accessed: 10 December 2011].

Kubo, K. 2009. *Democratisation of Former Yugoslavia and Nation Building Including Kosovo*. Paper Presented at the ISA's 50th Annual Convention, New York City, 15 February 2009. Available at: http://www.allacademic.com/meta/p313332_index.html [accessed: 27 November 2011].

Kuperman, A.J. 2008. The Moral Hazard of Humanitarian Intervention: Lessons from the Balkans. *International Studies Quarterly*, 52(1), 49–80.

Malcolm, N. 1998. *Kosovo: A Short History*. New York, NY: New York University Press.

Mayer, M.Z. 2005. The Eastern Question Unresolved: The US, and the Western Balkans. *International Journal*, 60(1), 237–46.

Mertus, J.A. 1999. *Kosovo: How Myths and Truths Started a War*. Berkeley, CA: University of California Press.

Narten, J. 2009. Assessing Kosovo's Postwar Democratisation: Between External Imposition and Local Self-Government. *Taiwan Journal of Democracy*, 5(1), 127–62.

Novick, P. 1988. *That Noble Dream: The Objectivity Question and the American Historical Profession*. Cambridge: Cambridge University Press.

Obradović, J. 2009. *Opinion – Kosovo: One Year on*. Paris: EU Institute for Security Studies.

OSCE 2000. *Municipal Elections 2000: Final Results*. Available at: http://www.osce.org/kosovo/20464 [accessed: 10 December 2011].

OSCE 2001. *Elections 2001: Certified Results*. Available at: http://www.osce.org/kosovo/20466 [accessed: 10 December 2011].

OSCE 2004. Elections 2004. Available at: http://www.osce.org/kosovo/25332 [accessed: 10 December 2011].

OSCE 2007. Election Results 2007. Available at: http://www.osce.org/kosovo/38260 [accessed: 10 December 2011].

Pajić, Z. 1995. Bosnia and Herzegovina: From Multiethnic Coexistence to "Apartheid" and Back, in *Yugoslavia: The Former and the Future. Reflections by Scholars from the Region*, edited by P. Akhavan and R. Howse. Washington, DC: The Brookings Institution, 152–63.

Papadimitriou, D. 2002. *Negotiating the New Europe: The European Union and Eastern Europe*. Aldershot: Ashgate.

Rambouillet Agreement 1999. Available at: http://jurist.law.pitt.edu/ramb.htm [accessed: 1 December 2011].

Ridder, E. De, Schrijvers, A. and Vos, H. 2010. Civilian Power Europe and Eastern Enlargement: The More the Merrier?, in *Europe's Global Role: External Policies of the European Union*, edited by J. Orbie. Farnham: Ashgate, 239–57.

Rosecrance, R. 1998. The EU: A New Type of International Actor, in *Paradoxes of European Foreign Policy*, edited by J. Zielonka. The Hague: Kluwer Law International, 15–24.

Sedelmeier, U. 2000. Eastern Enlargement: Risks, Rationality, and Role-Compliance, in *Risks, Reforms, Resistance and Revival: The State of the EU*, edited by M.G. Cowles and M. Smith. Oxford: Oxford University Press, 164–85.

Shala, B. 2003. *Vitet e Kosovës: 1998–1999*. Priština: Zëri.

Sjursen, H. 2002. Why Expand? The Question of Legitimacy and Justification in the EU's Enlargement Policy. *Journal of Common Market Studies*, 40(3), 491–513.

Skalnik Leff, C. 1999. Democratisation and Disintegration in Multinational States: The Breakup of Communist Federations. *World Politics*, 51(2), 205–35.

Stanchev, K. 2004. Economic Perspectives on Organised Crime. *Southeast European and Black Sea Studies*, 4(2), 241–9.

UN 1999. *UNSC Resolution 1244*. Available at: http://www.un.org/Docs/scres/1999/sc99.htm [accessed: 1 December 2011].

UNODC 2008. *Crime and its Impact on the Balkans and Affected Countries*. Available at: http://www.unodc.org/documents/Balkan_study.pdf [accessed 10 December 2011].

Vickers, M. 1994. *The Status of Kosovo in Socialist Yugoslavia*. Bradford: University of Bradford.

Weiss, G.T. and Kalbacher, Z.D. 2008. The United Nations, in *Security Studies: An Introduction*, edited by P.D. Williams. London: Routledge, 325–42.

Whitman, R. 2002. *The Fall and Rise of Civilian Power Europe*. The European Union in International Affairs, National Europe Centre, Australian National University Conference, 3–4 July 2002. Available at: https://digitalcollections.anu.edu.au/bitstream/1885/41589/3/whitman.pdf [accessed: 28 February 2012].

Chapter 9
The Future of the Post-Yugoslav Space

Branislav Radeljić

There is no doubt that it is difficult to talk about and especially to predict the future of the post-Yugoslav space. Such a difficulty has to do with a number of existing complex issues as well as numerous issues that are likely to come to occupy significant space in Balkan and European Union politics. In the early 1990s, the Europeans were expected to provide a solution to the Yugoslav state crisis, and their decisions resulted in the demise of the Yugoslav federation in January 1992. Later, the fighting in Croatia, Bosnia and Herzegovina and Kosovo intensified the presence of the EU in the region, which some welcomed and described as necessary while others perceived it as highly problematic.

However, over the last decade, the role of the EU has shifted from interventionist to integrationist, with a clear ambition to assist the post-Yugoslav states on their way towards EU membership. In this endeavour, as pointed out throughout the volume, the fulfilment of various criteria has been presented as an imperative capable of generating stability and prosperity in the entire region. Accordingly, this concluding chapter identifies and elaborates on some of the dominant concerns that could further erode the already asymmetrical progress of the post-Yugoslav states.

In, Almost In and Out

Slovenia, the most prosperous republic of the Yugoslav federation, managed to avoid much of the drama that accompanied the collapse of the Yugoslav state. In addition, the awareness of the moment, and the ability to implement reforms and convince the Brussels administration of its commitment resulted in Slovenia's accession to the EU in 2004. As some authors correctly observed, 'Slovenia gradually found ways to reconcile tension between the old and the new. As a result, both EU membership and NATO accession were gained because Slovenia had looked at itself and had begun to produce answers from within. In doing so, it helped to preserve the peculiarly Slovene while embracing change, in some ways integrating *slovenstvo* and cosmopolitan openness' (Gow and Carmichael 2010: 238).

Given Slovenia's progress, the country has been presented as a success story and, more relevantly, a model for others in the region. With regard to Croatia, the second most prosperous of the former Yugoslav republics, although heavily involved in the war of independence, it is possible to say that the EU managed to assist the Croatian structures in overcoming numerous obstacles, culminating in the likelihood of Croatia being the second post-Yugoslav state to join the EU. Indeed, the post-2000 period has been characterized by remarkable progress, mostly inspired by the local awareness of the benefits of eventual membership in the EU. In January 2012 the Croatian voters clearly supported their country's entry into the EU, seeing it as an opportunity to improve domestic progress and international standing.

Apart from Slovenia and Croatia, the European future of the remaining states is highly questionable. As often pointed out, Macedonia and Montenegro are expected to work harder in order to satisfy the membership criteria, as both are characterized by serious internal problems and divisions. With regard to Macedonia, although it secured candidacy for EU membership in 2005, there are still numerous areas that require significant improvements, including the fight against corruption, improvement of social policy and employment, promotion of competition policy, and so on. Even if there has been some progress, official assessments have often warned about various interruptions in the country's commitment towards reforms and thus EU membership. Given that such interruptions are closely linked to the interests and performance of the current Macedonian government, often criticized for placing too much emphasis on domestic policy and not enough on foreign policy and eventual benefits, it is understandable why it is difficult to talk about the country's European path. With regard to Montenegro, the Brussels administration supported its decision to become independent in 2006, but it also underlined that various sectors were in need of substantial improvements, making it clear that EU membership does not and will never go hand in hand with galloping corruption, organized crime or lack of an independent judiciary. The EU's official reports acknowledged that the country was trying to make some progress, although it was very limited. Once various areas had been identified as priorities and the EU, strongly believing in the attractiveness of EU membership, nevertheless decided to grant candidacy to Montenegro in 2010, the Montenegrin authorities almost immediately adopted a plan in order to address the outlined priorities, a strategy that resulted in the opening of accession negotiations in June 2012. Although it is difficult to say how much time it will take Montenegro to become a full EU member, it is possible to argue that, in contrast to some other (potential) EU candidates in the region, the Montenegrin authorities have had a clearer vision about their objectives, largely initiated by the attainment of independence.

Serbia, Bosnia and Herzegovina and Kosovo are more complex cases, mainly because of their existing interconnectedness, meaning that events and consequent policies occurring in one of them could influence developments in the other two. For example, with regard to the Kosovo case, the current situation, characterized by unviable political, economic and social components, could easily generate new violence, both within the borders of Kosovo and in its neighbouring countries. With this in mind, it is possible to argue that Kosovo is a Serbian, regional and European problem, all at the same time.

When discussing Kosovo as a Serbian problem, it is important to note that the Belgrade authorities have supported the Kosovo Serbs to remain in Kosovo. However, one of the most alarming issues concerns the presence in the province of Serbian enclaves that are still politically and economically integrated with Serbia. In the enclaves no one holds a stable job and the communities depend heavily on aid provided by Serbia. Moreover, while frustration is a dominant feature, the enclaves invite violence in order to level the playing field in society. Thus, how are these enclaves going to integrate within an independent Kosovo and who bears responsibility for this process? While the Serbian responsibility is to encourage Serbs willing to remain in Kosovo to work towards greater inclusion and representation in the society, instead of following what Belgrade officials have often suggested (which usually marginalizes them even more), what and how effective are Albanian and European responsibilities?

When seeing Kosovo as a regional problem, there are three main aspects to consider. First, the Albanian minority in Serbia, accounting for over 60,000 people, and predominantly inhabiting the southern part of the republic (Statistical Office of the Republic of Serbia 2003), may advocate autonomy. History suggests that young nations inhabiting a relatively small and limited territory but with high demographic growth, like the Albanians, tend to augment and claim the neighbouring territory, through war, migration and then secession. If the neighbouring nation, living in a comparatively large territory, is tired and relatively old, with low demographic growth, like the Serbs, the situation is ripe for instability in the form of claims to possession and territory. The second aspect concerns the coexistence of ethnic Macedonians and ethnic Albanians in Macedonia. The gap between these two peoples is alarming, considering that the Albanians are not satisfied with their status within the state. Such claims have a direct impact upon Macedonia's long-term stability. Back in 2000, the International Crisis Group warned that the two ethnic groups 'lead very separate and distinct lives' and that political leadership on both sides, although occasionally trying to talk, 'publicly cater to the more extreme nationalists in their respective parties' (ICG 2000). The situation has not improved much since then. Albanians in Macedonia tend to claim more rights, indeed, they often argue that Macedonians do not give them equality and have not implemented what they fought for and what was consequently promised after Macedonia had been dragged from the brink of civil war in 2001. According to one author, independent Kosovo 'promises to stabilize Macedonia by forestalling the radicalization of its ethnic Albanians and neutralizing Albanian extremists throughout the region', but in the end it is Macedonian responsibility and behaviour towards its Albanian minority 'that will do more to stabilize (or destabilize) the country than developments elsewhere' (Kupchan 2005). The third aspect concerns the Serbs inhabiting the artificially divided Bosnia and Herzegovina, who have not excluded the possibility of demanding secession since the independence of Kosovo became a reality. Thinking in such a direction is dangerous, as it could challenge both the 1995 Dayton-created peace and the stability of the wider Balkan region. As a response to Kosovo's independence, Branislav Dukić, leader of the Serb Movement of Independent Associations, called for Milorad Dodik, the Prime Minister of the Serb Republic, to include a referendum on independence in his electoral campaign. In his view, the Bosnian Serbs 'shall request independence for the Serb Republic as well. If Kosovo's illegal parliament can proclaim independence, the Bosnian Serb legal parliament should immediately proclaim independence for Republika Srpska even without calling for a referendum' (Dukić cited in Alić 2008). Needless to say, dismembering Bosnia and Herzegovina could possibly mean going back to the period of the Yugoslav state crisis.

When thinking about Kosovo as a European problem, there are also various aspects that deserve attention. First, given the numerous criteria that any country seeking membership in the European Union must satisfy, it is objective to say that Kosovo is likely to face numerous obstacles. Even when negotiating Kosovo's status, Brussels officials insisted on the policy of 'standards before status', inaugurated by the chief of the third UN Mission in Kosovo, Michael Steiner of Germany, but once it had become obvious that standards were not going to be fulfilled any time soon, the policy was abandoned. Second, Kosovo is formally recognized by 22 out of 27 Member States of the European Union. At present, the remaining five states (Cyprus, Greece, Romania, Slovakia and Spain) do not intend to recognize Kosovo's independence, as their decision to do so could generate various problems at home. While the existing scholarly explanations tend to approach such a decision from a broader perspective and thus treat it with more respect, some EU politicians see it as inappropriate. For example, Ulrike Lunacek, a Member of the European Parliament for Austria, has maintained that the reaction of the five non-recognition states is quite problematic as it also affects the others: '[W]ith only a few exceptions the 22 recognisers are leaning back, watching and waiting – rather than being active themselves in pushing for clarity and recognition. This weakens the EU's own efforts and diminishes the impact of the billions of euro being spent' (Lunacek 2012: 156). The third aspect concerns the possibility of partitioning Kosovo. While some officials, including Lunacek, argue that '[p]artition is out of the question' (Lunacek 2012: 152), some others see it as the only way forward. As observed by Sir Ivor Roberts, the former British Ambassador to Yugoslavia, 'less than a half of the UN members have recognised [Kosovo] as an independent state, and therefore it would be logical to make a new attempt to find a solution' and 'the partition of Kosovo would be the most logical solution' (*Tanjug* 2012a). Thinking further, the idea of partition could also be seen as the easiest way to obtain Serbia's consent to Kosovo's independence. Partition would probably please the Serbian public opinion as well as the Serbs living in the enclaves in independent Kosovo, although they would have to move to the northern part of the province or to Serbia proper.

The last point with regard to Kosovo as a European problem can be associated with the EU Member States' double standard on the presence of Islam in Europe. Although some of them tend to express sincere concerns about the integration of their Muslim minorities, they have completely ignored the fact that independent Kosovo is a Muslim state. Some recent events and consequent statements of the founding states of the European integrationist project have demonstrated that they struggle with their Muslim communities and concepts such as acculturation, assimilation and tolerance. However, none of them had any dilemmas in recognizing Kosovo within a fortnight of its proclamation of independence in February 2008. If analysed, the historical foundation of the European Union has been largely Christian-Democratic and, more importantly, the period around the 1992 Maastricht Treaty embraced Catholic social doctrine 'to guide intellectually and regulate institutionally cognitive meanings and political exigencies of a pluralist Europe' (Holmes 2009: 63). Thus, if the Europeans are concerned about the presence of Islam in their own countries, why are they so keen on the establishment of a Muslim state in the Western Balkans?

Apart from the Kosovo question, there are two other semi-acknowledged questions in Serbia, both capable of affecting European involvement and further integration of the country in the European Union. The first relates to the Serbian northern province of Vojvodina. As some authors put it, 'independence of Kosovo would have implications for Hungarians in Vojvodina' (Hadley 2008: 98). In 2006, the Office for European Affairs of the Autonomous Province of Vojvodina was established in Brussels with the aim to coordinate 'activities of provincial institutions in the European integration processes' (Executive Council of the Autonomous Province of Vojvodina 2006). The opening of the regional office does not necessarily indicate that Vojvodina will ignore Serbia (and the EU Integration Office of the Government of Republic of Serbia), but it does indicate that Vojvodina is ready to act on its own (based on its autonomous status) to foster its own links with the EU and secure benefits based on the level of commitment to the Union, regardless of the often-delayed policies of the Serbian government.

The second semi-acknowledged question relates to the Sandžak region, divided between Serbia and Montenegro. According to some observers, '[w]ith the Islamic Community holding "the balance of power in Sandžak", and manipulating its prominent position "to extract concessions from politicians and play a more active role in day-to-day politics" it is clear that political Islam can only grow stronger in the micro region' (Deliso 2007: 156). Mufti Muamer Zukorlić, head of the Islamic Community in Serbia, while supported by various external Muslim communities, has insisted on a greater autonomy for Sandžak that, 'would not challenge the sovereignty of either state, he says, nor would it be based on ethnicity' (*The Economist* 2011). But what if the already weakened dialogue between the Belgrade and Sandžak authorities deteriorates further, to the extent that independence of the Sandžak region becomes the dominant issue? Here, again, what role would the Europeans play?

Serbia is expected to give up Kosovo if it wants to become a full member of the European Union. Does this mean that if Vojvodina and Sandžak decide to claim independence from Serbia, the Europeans (or the West) will be there to assist them by making Serbia's progress towards EU membership conditional on their recognition? Such questions make even more sense when one considers that Serbia does not have an alternative to the European Union. Although the lack of alternative was confirmed during the Djindjić government, the leadership that took over has occasionally struggled to confirm this view, leaving an impression that Serbia might have an alternative elsewhere. The Russian Federation is not an option for Serbia, neither economically, nor ideologically. With regard to economics, the EU is Serbia's biggest trade partner, accounting for 56 per cent of domestic exports (*Tanjug* 2012a). With regard to ideology, although Titoist Yugoslavia was often proud of its similarity to the Soviet Union, the present-day Serbs are trying to be closer to the EU, and the West in general, than to the Russians. In addition, Russia's decision not to recognize Kosovo's independence has nothing to do with ideological proximity, but with its own ambition to be more relevant in the European Union and international politics.

A Common Denominator

The relevance of the European Union for the post-Yugoslav space and the benefit of having a stable Western Balkan region are not open to question. While in the 1990s the presence of the EU was interventionist in nature, in the 2000s and after it has been largely integrationist. For example, given the growing pressure to bring the Kosovo case to a solution, it is believed that it is only the EU that can facilitate the whole process, by bringing the Serbs and the Albanians closer together (despite the fact that the outcome will not manage to satisfy both sides) and by offering to both of them a clear framework for EU membership, in order to prevent a new conflict in the near future.

The Brussels administration has relied on its criteria and conditionality to democratize and Europeanize the states in the region so that they can eventually become full EU members. Various scholars are right when noting that '[o]ne of the most important lessons of the Western Balkan case is that the pace of democratic consolidation is mainly attributable to the willingness of postconflict elites to work together in tackling the entrenched sources of authoritarian resilience' (Cohen and Lampe 2011: 494). However, the recent crisis of the euro and consequent policies adopted across the EU have generated doubts about its enlargement agenda for South Eastern Europe and the expectations among (potential) EU candidates. As one study correctly suggests, '[f]or such doubts to be dispelled, two complementary things are needed. Firstly, a strong positive signal from Brussels, in the form of an accelerated and coherent EU regional expansion policy in the Balkans. Secondly, and no less importantly: the capacity of local actors to tackle the European reform agenda not just as something that is imposed from outside, but as domestic homework for any democratic European society in the twenty-first century' (Rupnik 2011: 28).

In the whole process, one is still more likely to get the impression that it has been the EU that is more interested in providing the Western Balkans with a European perspective than the respective states themselves. As often outlined, actors involved in transnational organized crime and monopolistic businesses, although seriously affecting the development and stability of the Balkans, have shown no interest whatsoever in European integration. Indeed, some of them have tried to establish close links with political elites, so that their position could be even more protected. As a consequence, such benefit-inspired relationships have represented an obstacle to legal, political and economic reforms. Back in 2005, one author assessed the situation in Serbia and stressed that '[p]olitical renewal in Serbia depends on a reconfiguration of domestic forces resulting in a full-fledged effort to restore the rule of law and root out corruption. But such a modernising strategy also depends on the success of wider European integration processes and the wisdom of specific European strategies towards Serbia. Fresh thinking about how to breathe new life into a flagging reform process was far from evident either at home or in the capitals of the European Union as moderates struggled to retain the initiative against emboldened ultra-nationalist forces' (Gallagher 2005: 131).

In the meantime, the situation has improved, but still not enough so that the Western Balkans should be applauded for its achievements. In fact, almost every single progress report has named corruption as a huge problem in the region, and thus the question of how successful the EU's involvement and requests are, given the local reluctance to fully cooperate with the Brussels administration, is a valid one. Accordingly, some studies are right when insisting on the relevance of transparency as an absolute prerequisite for democratic development and fulfilment of the EU political criteria. As one of them puts it, 'governments need "watchdog"-style assessors with powers both within a bureaucracy and across bureaucracies and other component parts of the political system' (Farkas 2007: 90). In addition, the connection between the governed and those who govern should go beyond the elections, as the continuum of the elite–mass interface encourages further transparency and accountability (Farkas 2007: 93). Thus, the public should be entitled to know who does what and when and have the right to express itself. Voice represents a concept whose importance primarily depends on a stable political architecture. Media, social movements and other nonpolitical relationships can act as facilitating factors to make the public voice heard properly. In addition, the role of civil society is 'to nurture grassroots political activity in Southeast Europe that could become the basis for widespread, genuine public participation' (Farkas 2007: 63). The list of concerns is further complemented by rather disappointing records in regard to tolerance. As observed in the literature, most societies in the Balkans 'have very little experience with tolerance' due to their 'actively intolerant political regimes' in the past – a feature that surely affects the speed of transition (Farkas 2007: 24). As the decision of the Serbian authorities to ban the 2012 Gay Pride in Belgrade has confirmed, difference generates intolerance.

Inevitably, the above noted issues pose new questions and encourage further analyses that, apart from having to do with the EU's political criteria, have a lot to do with local mentalities. However, what deserves the greatest recognition is the fact that the desire for eventual membership in the European Union has been a common denominator for all Western Balkan countries. It is an incentive to develop modern and effective legislation, and to reform and stabilize the political environment.

Needless to say, without full support from Brussels, there is little hope that the region will be able to shake off its reputation. The worst-case scenario would be that it allows an increase in the already evident disproportion between the region and its neighbours who are further advanced in the EU integration process. The Balkan question was and has remained a European question. Luckily, on various occasions, the Europeans have confirmed their commitment to the region. In this respect, we could agree and hope that the future represents an opportunity to break with the past, and while for some countries this entails taking responsibility for their acts in the past, it also, in turn, facilitates new beginnings.

References

Alić, A. 2008. *Bosnian Serbs Protest over Kosovo*. Available at: http://www.isn.ethz.ch/isn/Security-Watch/Articles/Detail/?ots591=4888caa0-b3db-1461-98b9-e20e7b9c13d4&lng=en&id=54209 [accessed: 12 August 2012].

Cohen, L.J. and Lampe, J.R. 2011. *Embracing Democracy in the Western Balkans: From Post-Communist Struggles toward European Integration*. Washington, DC: Woodrow Wilson Center Press.

Deliso, C. 2007. *The Coming Balkan Caliphate: The Threat of Radical Islam to Europe and the West*. Westport, CT: Praeger.

Economist, The 2011. Together We Are Stronger! Rabble-Rousing in a Muslim-Majority Part of Serbia. 27 August 2011.

Executive Council of the Autonomous Province of Vojvodina 2006. *Decision on Establishment of the Office on European Affairs*. Novi Sad, 5 April 2006.

Farkas, R.P. 2007. *Democratization in the Balkans: Prescription for a Badly Scarred Body Politic*. Boston, MA: Northeastern University Press.

Gallagher, T. 2005. *The Balkans in the New Millennium: In the Shadow of War and Peace*. Oxford: Routledge.

Gow, J. and Carmichael, C. 2010. *Slovenia and the Slovenes: A Small State in the New Europe*. London: Hurst and Company.

Hadley, J. 2008. The Way Opened, the Way Blocked: Assessing the Contrasting Fates of Chechnya and Kosovo, in *On the Way to Statehood: Secession and Globalization*, edited by A. Pavković and P. Radan. Aldershot: Ashgate, 85–99.

Holmes, D.R. 2009. Experimental Identities (after Maastricht), in *European Identity*, edited by J.T. Checkel and P.J. Katzenstein. Cambridge: Cambridge University Press, 52–80.

ICG 2000. *Report No. 98*. Available at: http://www.crisisgroup.org/home/index.cfm?l=1&id=1162&sr=8 [accessed: 20 July 2012].

Kupchan, C.A. 2005. Independence for Kosovo. *Foreign Affairs*. Available at: http://www.foreignaffairs.org/20051101facomment84603/charles-a-kupchan/independenceforkosovo.html [accessed: 10 July 2012].

Lunacek, U. 2012. Kosovo-Serbia: Will They Ever Come to Terms? What is the European Union's Role?, in *Unfinished Business: The Western Balkans and the*

International Community, edited by V. Džihić and D. Hamilton. Washington, DC: Center for Transatlantic Relations, 149–57.

Rupnik, J. 2011. The Balkans as a European Question, in *The Western Balkans and the EU: 'The Hour of Europe'*, edited by J. Rupnik. Paris: EU Institute for Security Studies, 17–30.

Statistical Office of the Republic of Serbia 2003. *Popis stanovništva, domaćinstava i stanova 2002; Knjiga 1: Nacionalna ili etnička pripadnost po naseljima*. Belgrade: Statistical Office of the Republic of Serbia.

Tanjug 2012a. Delević: EU Is Serbia's Main Trade Partner. Available at: http://www.tanjug.rs/news/36110/delevic--eu-is-serbias-main-trade-partner.htm [accessed: 19 August 2012].

Tanjug 2012b. Former British Ambassador Backs Kosovo Partition. Available at: http://www.tanjug.rs/news/38541/former-british-ambassador-backs-kosovo-partition.htm# [accessed: 8 September 2012].

Index

Page numbers in *italic* type refer to tables. Numbers followed by 'n' refer to footnotes.